Crystallography Made Crystal Clear

Crystallography Made Crystal Clear

A Guide for Users of Macromolecular Models

Gale Rhodes
Chemistry Department
University of Southern Maine
Portland, Maine

ACADEMIC PRESS, INC.
Harcourt Brace & Company
San Diego   New York   Boston   London   Sydney   Tokyo   Toronto

Cover photo: Cytochrome *b*5 with dotted extended surface on heme. See Plate 13 for further details.

Academic Press, Inc.
1250 Sixth Avenue, San Diego, California 92101-4311

United Kingdom Edition published by
Academic Press Limited
24–28 Oval Road, London NW1 7DX

Library of Congress Cataloging-in-Publication Data

Rhodes, Gale.
    Crystallography made crystal clear : a guide for users of
macromolecular models / Gale Rhodes.
        p. cm.
    Includes index.
    ISBN 0-12-587075-2
    1. Proteins--Analysis. 2. X-ray crystallography. I. Title.
QP551.R48   1993
547.7'5046--dc20   92-43102
                                                           CIP

PRINTED IN THE UNITED STATES OF AMERICA
93 94 95 96 97 98   MM   9 8 7 6 5 4 3 2 1

To Pam
(THE h.p.f.w.a.t.r.a.n.Z!)
and to the memory of
Carl and Eugenia Rhodes

Contents

Preface   xi

1. Model and Molecule   1

2. An Overview of Protein Crystallography   5

Introduction   5
Crystals   8
Collecting X-ray Data   10
Diffraction   12
Coordinate Systems in Crystallography   18
The Mathematics of Crystallography: A Brief Description   19

3. Protein Crystals   29

Properties of Protein Crystals   29
Evidence That Solution and Crystal Structures Are Similar   33
Growing Protein Crystals   35
Judging Crystal Quality   40

4. Collecting Diffraction Data   43

Introduction   43
Geometric Principles of Diffraction   43
Collecting X-ray Diffraction Data   61
Summary   81

5. From Diffraction Data to Electron Density   83

Introduction   83
Fourier Series and the Fourier Transform   84
Fourier Mathematics and Diffraction   91
The Meaning of the Fourier Equations   94
Summary: From Data to Density   99

6. Obtaining Phases   101

Introduction   101
Two-Dimensional Representation of Structure Factors   102
The Heavy-Atom Method (Isomorphous Replacement)   107
Anomalous Scattering   118
Molecular Replacement: Related Proteins as Phasing Models   125
Iterative Improvement of Phases (Preview of Chapter 7)   129

7. Obtaining and Judging the Molecular Model   131

Introduction   131
Iterative Improvement of Maps and Models: Overview   131
First Maps   134
The Model Becomes Molecular   139
Structure Refinement   144
Convergence to a Final Structure   149
Sharing the Model   153

8. A User's Guide to Crystallographic Models   157

Introduction   157
Judging the Quality and Usefulness of the Refined Model   158
Reading a Crystallography Paper   166
Summary   183

9. Tools for Studying Proteins   185

Introduction   185
Computer Models of Molecules   185

Touring a Typical Molecular Modeling Program   189
Other Tools for Studying Structure   197
A Final Note   198

Index   199

Preface

Most texts that treat biochemistry or proteins contain a brief section or chapter on protein crystallography. Even the best of such sections are usually mystifying — far too abbreviated to give any real understanding. In a few pages, the author can accomplish little more than telling you to have faith in the method. At the other extreme are many useful treatises for the would-be, novice, or experienced crystallographer. Such accounts contain all the theoretical and experimental details that practitioners must master and, for this reason, they are quite intimidating to the noncrystallographer. This book lies in the vast and heretofore empty region between brief textbook sections and complete treatments of the method aimed at the professional crystallographer. I hope there is just enough here to help the noncrystallographer understand where crystallographic models come from, how to judge their quality, and how to glean additional information that is not depicted in the model but *is* available from the crystallographic study that produced the model.

This book should be useful to protein researchers in all areas; to students of biochemistry in general and of macromolecules in particular; to teachers as an auxiliary text for courses in biochemistry, biophysical methods, and macromolecules; and to anyone who wants an intellectually satisfying understanding of how crystallographers obtain three-dimensional models of macromolecules. This understanding is essential for intelligent use of crystallographic models, whether that use is the study of molecular action and interaction, an attempt to unlock the secrets of protein folding, exploration of the possibilities of engineering new protein functions, or interpretation of the results of chemical, kinetic, thermodynamic, or spectroscopic experiments on proteins. Indeed, if you use models without knowing how they were obtained, you may be treading on hazardous ground. For instance, you may fail to use available information that would give you greater insight into the molecule and its action. Or worse, you may devise and publish a detailed molecular explanation based on a structural feature that is quite uncertain. Fuller understanding of the strengths and limitations of crystallographic models will enable you to use them wisely and effectively.

If you are part of my intended audience, I do not believe you need to know, or are likely to care about, all the gory details of crystallographic methods and all the esoterica of crystallographic theory. I present just enough about methods to give you a feeling for the experiments that produce crystallographic data. I present somewhat more theory, because it underpins an understanding of the nature of a crystallographic model. I want to help you follow a logical thread that begins with diffraction data and ends with a colorful picture of a protein model on the screen of a graphics computer. The novice crystallographer, or the student pondering a career in crystallography, may find this book a good place to start, a means of seeing if the subject remains interesting under closer scrutiny. But these readers will need to consult more extensive works for fine details of theory and method. I hope that reading this book makes those texts more accessible.

(I assume that you are familiar with protein structure, at least at the level presented in an introductory biochemistry text. My own favorite treatment of this subject for beginners is Chapters 3 through 5 of David Rawn's *Biochemistry*, Neil Patterson Publishers, 1989.)

I wish I could teach you about crystallography without using mathematics, simply because so many readers are apt to throw in the towel upon turning the page and finding themselves confronted with equations. Alas (or hurrah, depending on your mathematical bent), the real beauty of crystallography lies in the mathematical and geometric relationships between diffraction data and molecular images. I attempt to resolve this dilemma by presenting no more math than is essential and taking the time to *explain in words what the equations imply*. Where possible, I emphasize geometric explanations over equations.

If you turn casually to the middle of this book, you will see some forbidding mathematical formulae. Let me assure you that I move to those bushy statements step by step from nearby clearings, making minimum assumptions about your facility and experience with math. For example, when I introduce periodic functions, I tell you how the simplest of such functions (sines and cosines) "work," and then I move slowly from that clear trailhead into the thicker forest of complicated wave equations that describe x-rays and the molecules that diffract them. When I first use complex numbers, I define them and illustrate their simplest uses and representations, sort of like breaking out camping gear in the dry safety of a garage. Then I move out into real weather and set up a working camp, showing how the geometry of complex numbers reveals essential information otherwise hidden in the data. My goal is to help you see the relationships implied by the mathematics, not to make you a calculating athlete. My ultimate aim is to prove to you that the structure of molecules really does lie lurking in the crystallographic data—that, in fact, the information in the diffraction pattern implies a unique structure. I hope thereby to remove the mystery about how structures are coaxed from data.

If, in spite of these efforts, you find yourself flagging in the most technical chapters (4 through 7), please do not quit. I believe you can follow the arguments of these chapters and thus be ready for the take-home lessons of Chapters 8 and 9, even if the equations do not speak clearly to you. Jacob Bronowski once described the verbal argument in mathematical writing as analogous to melody in music, and thus a source of satisfaction in itself. He likened the equations to musical accompaniment that becomes more satisfying with repeated listening. If you follow and retain the melody of arguments and illustrations in Chapters 4 through 7, then the last chapters and their take-home lessons should be useful to you.

I aim further to enable you to read primary journal articles that announce and present new protein structures, including the arcane sections on experimental methods. In most scientific papers, experimental sections are directed primarily toward those who might use the same methods. In crystallographic papers, however, methods sections contain information from which the quality of the model can be roughly judged. This judgment should affect your decision about whether to obtain the model and use it, and whether it is good enough to serve as a guide in drawing the kinds of conclusions you hope to draw. In Chapter 8, to review many concepts, as well as to exercise your new skills, I look at and interpret experimental details in literature reports of a recent structure determination.

Finally, I hope you read this book for pleasure — the sheer pleasure of turning the formerly incomprehensible into the familiar. In a sense, I am attempting to share with you my own pleasure of the past ten years, after my mid-career decision to set aside other interests and finally see how crystallographers produce the molecular models that have been the greatest delight of my teaching. Among those I should thank for opening their labs and giving their time to an old dog trying to learn new tricks are Professors Leonard J. Banaszak, Jens Birktoft, Jeffrey Bolin, John Johnson, and Michael Rossmann.

I would never have completed this book without the patience of my wife, Pam, who allowed me to turn part of our home into a miniature publishing company, nor without the generosity of my faculty colleagues, who allowed me a sabbatical leave during times of great economic stress at the University of Southern Maine. Many thanks to Lorraine Lica, my Acquisitions Editor at Academic Press, who grasped the spirit of this little project from the very beginning and then held me and a full corps of editors, designers, and production workers accountable to that spirit throughout.

Gale Rhodes

1 Model and Molecule

Proteins perform many functions in living organisms. For example, some proteins regulate the expression of genes. One class of gene-regulating proteins contains structures known as "zinc fingers," which bind directly to DNA. Plate 1 shows a complex composed of a double stranded DNA (deoxyribonucleic acid) molecule and three zinc fingers from the mouse protein Zif268.

The protein backbone is shown in yellow and all side chains in blue. The two DNA strands are red and green. Three zinc atoms, which are complexed to side chains in the protein, are purple. The yellow dotted line indicates a hydrogen bond in which a nitrogen atom of arginine-18 (in the protein) and a nitrogen atom of guanine-10 (in the DNA) share a hydrogen atom, an interaction that holds the nitrogens 2.79 Å apart. If you look closely at the photograph, you can see that all of the protein–DNA interactions are between protein side chains and DNA bases; the protein backbone does not come in contact with the DNA. Looking more closely at the photo, or studying this complex on a modern graphics computer, you could discover the specific interactions between side chains of Zif268 and base pairs of DNA. You could enumerate the hydrogen bonds and other contacts that stabilize this complex and cause Zif268 to recognize a specific sequence of bases in DNA. You might gain some testable insights into how the protein finds the correct DNA sequence amid the vast amount of DNA in the nucleus of a cell. The structure might also lead you to speculate on how alterations in the sequence of amino acids in the protein might result in affinity for different DNA sequences, and thus start you thinking about how to design other DNA-binding proteins.

Now look again at the preceding paragraph and examine its language rather than its content. The language is typical of that in common use to describe molecular structure and interactions as revealed by various experimental methods, including single-crystal x-ray crystallography, the subject of this book. In fact, this language is shorthand for more precise but cumbersome statements of what we learn from structural studies. First, Plate 1 of course shows not molecules but *models* of molecules, in which structures and interactions are *depicted*, not shown. Second, in this specific case, the models are of molecules not in solution but in the crystalline state, because the models are derived from analysis of x-ray diffraction by crystals of the Zif268/DNA complex. As such, these models depict the average structure of somewhere between 10^{13} and 10^{15} complexes throughout the crystals that were studied. In addition, the structures are averaged over the time of the x-ray experiment, which is at least several days.

Drawing the conclusions of the first paragraph requires bringing additional knowledge to bear on the graphics image, including knowledge of just what we learn from x-ray analysis. (The same could be said for structural models derived from spectroscopic data or any other method.) In short, the graphics image itself is incomplete. It does not reveal things we may know about the complex from other types of experiments, and it does not even reveal all that we learn from x-ray crystallography.

For example, how accurately are the relative positions of atoms known? Are the nitrogen atoms of arginine-18 and guanine-10 precisely 2.79 Å apart, or is there some tolerance in that figure? Is the tolerance large enough to jeopardize the conclusion that a hydrogen bond joins these atoms? Further, do we know anything about how rigid this complex is? Do parts of these molecules vibrate, or do they move with respect to each other? Still further, in the aqueous medium of the cell, does this complex have the same structure as in the crystal, which is a solid? As we examine this model, are we really gaining insight into cellular processes? A final question may surprise you: Does the model fully account for the chemical composition of the crystal? In other words, are any of the known contents of the crystal missing from the model?

The answers to these questions are not revealed in the graphics image, which is more akin to a cartoon than to a molecule. Actually, the answers vary from one model to the next, but they are usually available to the user of crystallographic models. Some of the answers come from x-ray crystallography itself, so the crystallographer does not miss or overlook them; they are simply less accessible to the noncrystallographer than is the graphics image.

Viewing Stereo Images

To see a three-dimensional image of these models, use a stereo viewer such as item #46-9000, Carolina Biological Supply Company, PO Drawer 2827, Burlington, NC 27216-2827. You can view stereo pairs without a viewer by training yourself to look at the left image with your left eye and the right image with your right eye. This is neither as difficult nor as strange as it sounds. (According to my ophthalmologist, it is not harmful to the eyes, and may in fact be good exercise for eye muscles.) Try putting your nose on the page between the two views. With both eyes open, you will see the two images superimposed but out of focus, because they are too close to your eyes. Slowly move the paper away from your face, trying to keep the images superimposed until you can focus on them. (Keep the line between image centers parallel to the line between your eyes.) When you can focus, you will see three images. The middle one should exhibit convincing depth. Try to ignore the flat images on either side. This process becomes easier with practice. You may find it helpful to try this process first on one of the simpler images, such as Plate 5 or Plate 12.

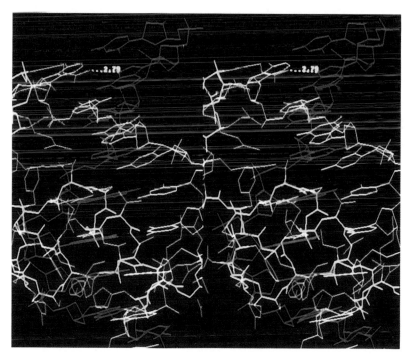

Plate 1 Stereo photograph of Zif268/DNA complex [see N. P. Pavletich and C. O. Pabo, *Science* **252**, 809 (1991)]. Atomic coordinates generously provided by N. P. Pavletich. (For discussion see Chapter 1.)

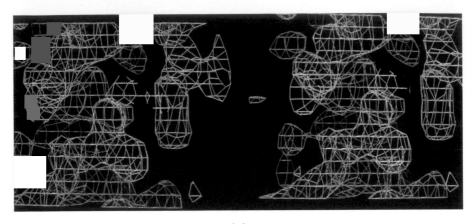

(a)

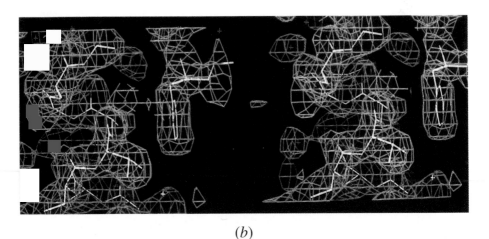

(b)

Plate 2 (*a*) Small section of a molecular image displayed on a computer graphics terminal. (*b*) Image (*a*) is *interpreted* by building a molecular model within the image. Computer graphics programs allow parts of the model to be added and their conformations adjusted to fit the image. The protein shown here is adipocyte lipid-binding protein (ALBP). Atomic coordinates courtesy of Professor Leonard J. Banaszak. (For discussion see Chapter 2.)

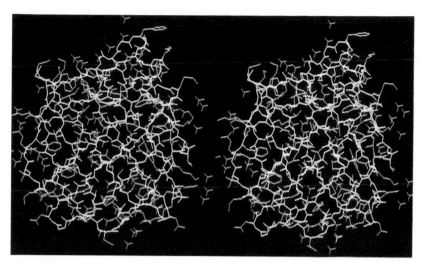

Plate 3 One molecule of crystalline adipocyte lipid-binding protein (ALBP), showing ordered water molecules on the surface and within a molecular cavity where lipids are usually bound. Protein backbone is yellow, side chains are blue, and ordered water molecules are green. (For discussion see Chapter 3.)

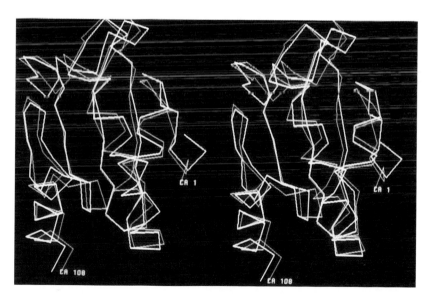

Plate 4 Structures of the *E. coli* protein thioredoxin as determined by x-ray crystallography (white) and by NMR (green). Only backbone α-carbons are shown. The models were superimposed by least-squares minimization of the distances between corresponding atoms. Atomic coordinates obtained from the Protein Data Bank, which is described in Chapter 7. (For discussion see Chapter 3.)

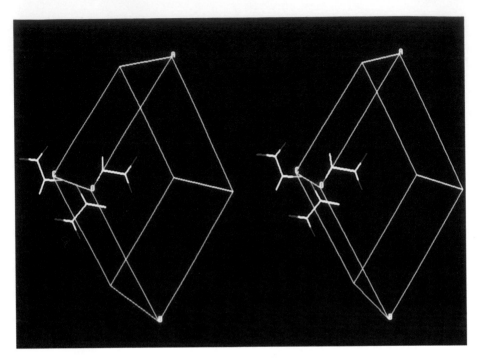

Plate 5 Threefold screw axis (3₁). (For discussion see Chapter 4.)

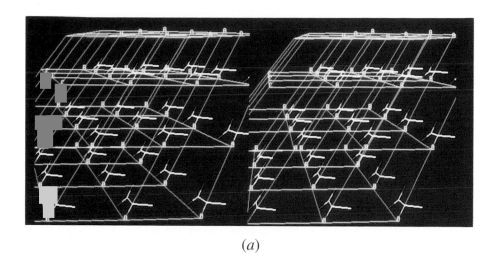

(a)

(b)

Plate 6 Alanine in hypothetical (a) $P1$ and (b) $P2_1$ unit cells. (For discussion see Chapter 4.)

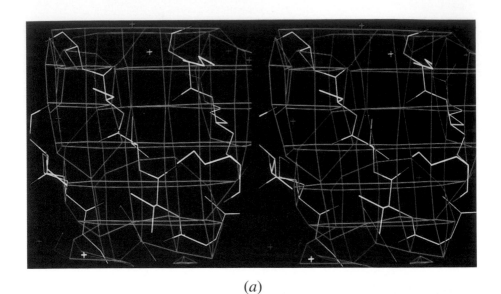

(*a*)

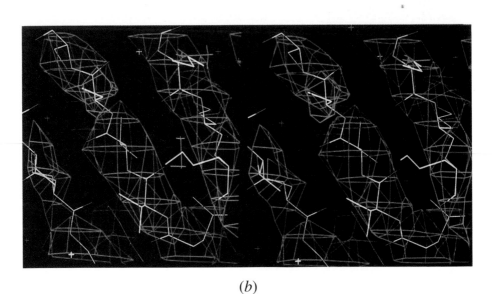

(*b*)

Plate 7 Electron-density maps at increasing resolution. Maps were calculated using final phases, and Fourier series were truncated at the resolution limits indicated: (*a*) 6.0 Å, (*b*) 4.5 Å, (*c*) 3.0 Å, (*d*) 1.6 Å. (For discussion see Chapter 7.)

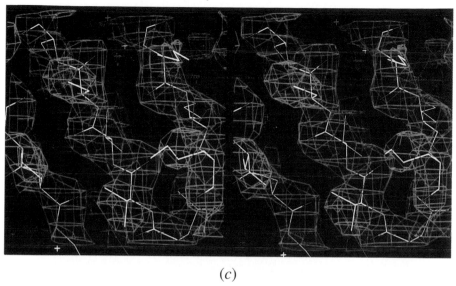

(c)

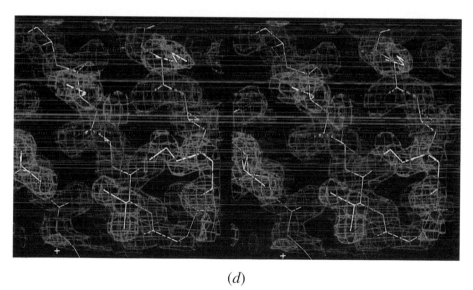

(d)

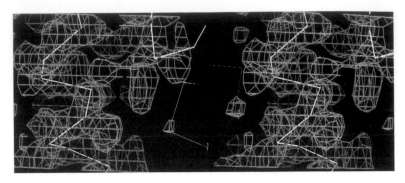

Plate 8 α-Carbon model of ALBP built into an electron-density map. (For discussion see Chapter 7.)

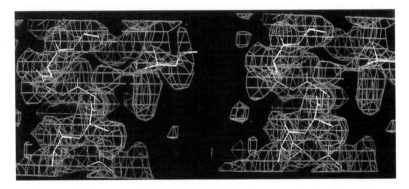

Plate 9 Polyalanine model of ALBP built into an electron-density map. This section of the final ALBP model is shown in Plate 2. (For discussion see Chapter 8.)

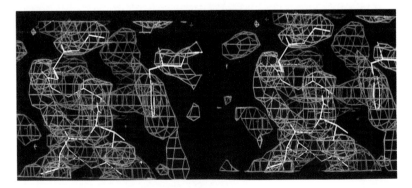

Plate 10 ALBP electron-density map calculated with molecular-replacement phases before any refinement, shown with the final model. Compare with Plate 2, which shows the final electron-density map in the same region. (For discussion see Chapter 8.)

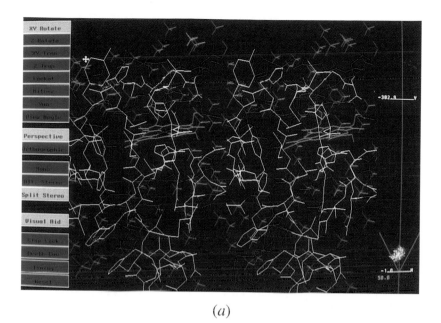

(a)

(b)

Plate 11 (a) The heme area of cytochrome b5, showing all atoms, without clipping. (b) Same view with z-clipping to remove foreground and background atoms. In both views, line of sight, viewing angle, and clipping planes are depicted on the right side of the screen. (For discussion see Chapter 9.)

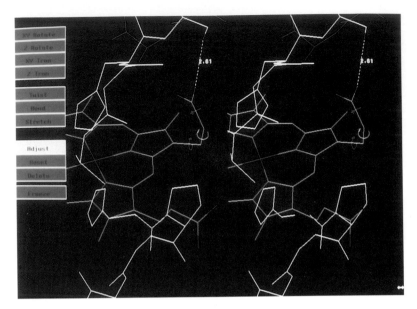

Plate 12 Measurements defining a hydrogen bond (yellow) and bond rotation in progress (curved arrows). (For discussion see Chapter 9.)

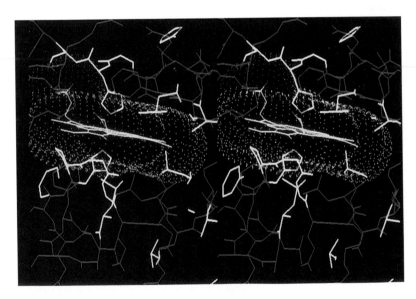

Plate 13 Cytochrome *b*5 with dotted extended surface on heme. Protein atoms in contact with this surface are two atom diameters from centers of heme atoms, and thus they are considered to be in contact with the heme. (For discussion see Chapter 9.)

   Molecular models obtained from crystallography are in wide use as tools
for revealing molecular details of life processes. Scientists use models to
learn how molecules "work": how enzymes catalyze metabolic reactions,
how transport proteins load and unload their molecular cargo, how antibod-
ies bind and destroy foreign substances, how proteins turn genes on and off.
It is easy for the user of crystallographic models, being anxious to turn otherwise puzzling information into a mechanism of action, to treat models as
everyday objects seen as we see clouds, birds, and trees. But the informed
user of models sees more than the graphics image, recognizing it as a static
depiction of dynamic objects, as the average of many similar structures, as
perhaps lacking parts that are present in the crystal but not revealed by the
x-ray analysis, and finally as a fallible interpretation of data. The informed
user knows that the crystallographic model is richer than the cartoon.

   In the following chapters, I offer you the opportunity to become an informed user of crystallographic models. Knowing the richness and limitations of models requires understanding the relationship between data and
structure. In Chapter 2, I give an overview of this relationship. In Chapters
3 through 7, I simply expand Chapter 2 in enough detail to produce an intact
chain of logic stretching from diffraction data to the final model. Topics
come in roughly the same order as the tasks that face a crystallographer pursuing a molecular structure.

   As a practical matter, informed use of a model requires reading the crystallographic papers that report the new structure, and extracting from them
criteria of model quality. In Chapter 8, I discuss these criteria and provide
a guided exercise in extracting them The exercise takes the form of annotated excerpts from a recent structure determination. Equipped with the
background of previous chapters, and experienced with the real-world exercise of a guided tour through a recent publication, you should be able to
read new structure publications in the scientific literature, understand how
the structures were obtained, and be aware of just what is known, and what
is still unknown, about the molecules under study.

   For new or would-be users of models, I present in Chapter 9 a brief introduction to molecular modeling, demonstrating how modern graphics
computers and programs allow users to display and manipulate models.

   Today's scientific textbooks and journals are filled with stories about the
molecular processes of life. The central character in these stories is often a
protein molecule, a thing never seen in action, never perceived directly. We
see model molecules in books and on computer screens, and we tend to treat
them as everyday objects accessible to our normal perceptions. In fact,
models are hard-won products of technically difficult data collection and
powerful but subtle data analysis. This book concerns where our models of
structure come from, and how to use them wisely.

2 An Overview of Protein Crystallography

I. Introduction

The most common means of obtaining a detailed picture of a large molecule, allowing the resolution of individual atoms, is to interpret the diffraction of x-rays from many identical molecules in an ordered array like a crystal. This method is called *single-crystal x-ray crystallography.* As of this writing, roughly 1000 protein structures have been obtained by this method. Recently, the structures of a number of small proteins have been solved by nuclear magnetic resonance (NMR) spectroscopy, which provides a model of the protein in solution, rather than in the crystalline state. Both methods have their strengths and weaknesses, so they will undoubtedly coexist as complementary methods in the foreseeable future. One of the goals of this book is to make users of crystallographic models aware of the strengths and weaknesses of x-ray crystallography, so that users' expectations of the resulting models are in keeping with the limitations of crystallographic methods.

This chapter provides a simplified overview of how researchers use the technique of x-ray crystallography to learn macromolecular structures. Chapters 3–8 are simply expansions of the material in this chapter. I will

speak primarily of proteins, but the concepts I describe apply to all macromolecules and macromolecular assemblies that possess ordered structure,
including carbohydrates, nucleic acids, and nucleoprotein complexes like
ribosomes and whole viruses.

A. Obtaining an image of a microscopic object

When we see an object, light rays bounce off (are diffracted by) the object
and enter the eye through the lens, which reconstructs an image of the object and focuses it on the retina. In a simple microscope, an illuminated
object is placed just beyond one focal point of a lens, which is called the
objective lens. The lens collects light diffracted from the object and reconstructs an image beyond the focal point on the opposite side of the lens, as
shown in Fig. 2.1.

For a simple lens, the relationship of object position to image position in
Fig. 2.1 is $(OF)(IF') = (FL)(F'L)$. Because the distances FL and $F'L$ are
constants (but not necessarily equal) for a fixed lens, the distance OF is inversely proportional to the distance IF'. Placing the object near the focal
point F results in a magnified image produced at a considerable distance
from F' on the other side of the lens, which is convenient for viewing. In a
compound microscope, the most common type, an additional lens, the *eye-
piece*, is added to magnify the image produced by the objective lens.

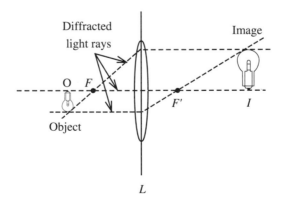

Figure 2.1 Action of a simple lens. Rays parallel to the lens axis strike the lens
and are refracted into paths passing through a focus. Rays passing through a focus
strike the lens and are refracted into paths parallel to the lens axis. As a result, the
lens produces an image at I of an object at O, such that $(OF)(IF') = (FL)(F'L)$.

B. Obtaining images of molecules

In order for the object to diffract light and thus be visible under magnification, the wavelength (λ) of the light must be, roughly speaking, no larger than the object. Visible light, which is electromagnetic radiation with wavelengths of 400–700 nm (1 nm = 10^{-9} m), cannot produce an image of individual atoms in protein molecules, in which bonded atoms are only about 0.15 nm or 1.5 Å (Å = 10^{-10} m) apart. Electromagnetic radiation of this wavelength falls into the x-ray range, so x-rays are diffracted by even the smallest molecules. X-ray analysis of proteins seldom resolves the hydrogen atoms, so the protein models described in this book include elements on only the second and higher rows of the periodic table. The positions of all hydrogen atoms can be deduced on the assumption that bond lengths, bond angles, and conformational angles in proteins are just like those in small organic molecules.

Even though individual atoms diffract x-rays, it is still not possible to produce a focused image of a molecule, for two reasons. First, x-rays cannot be focused by lenses. Crystallographers sidestep this problem by measuring the directions and strengths (intensities) of the diffracted x-rays and then using a computer to simulate an image-reconstructing lens. In short, the computer acts as the lens, computing the image of the object and then displaying it on a screen or drawing it on paper (Fig. 2.2).

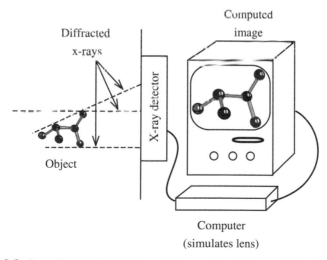

Figure 2.2 Crystallographic analogy of lens action. X-rays diffracted from the object are received and measured by a detector. The measurements are fed to a computer, which simulates the action of a lens to produce a graphics image of the object.

Second, a single molecule is a very weak diffractor of x-rays. Most of the x-rays will pass through a single molecule without being diffracted, so the diffracted beams are too weak to be detected. Analyzing diffraction from crystals, rather than individual molecules, solves this problem. A crystal of a protein contains many ordered molecules in identical orientations, so each molecule diffracts identically, and the diffracted beams for all molecules augment each other to produce strong, detectable x-ray beams.

C. A thumbnail sketch of protein crystallography

In brief, determining the structure of a protein by x-ray crystallography entails growing high-quality crystals of the purified protein, measuring the directions and intensities of x-ray beams diffracted from the crystals, and using a computer to simulate the effects of an objective lens and thus produce an image of the crystal's contents, like the small section of a molecular image shown in Plate 2a. Finally, that image must be interpreted, which entails displaying it by computer graphics and building a molecular model that is consistent with the image (Plate 2b).

The resulting model is often the only product of crystallography that the user sees. It is therefore easy to think of the model as a real entity that has been directly observed. In fact, our "view" of the molecule is quite indirect. Understanding just how the crystallographer obtains models of protein molecules from diffraction measurements is essential to fully understanding how to use models properly.

II. Crystals

A. The nature of crystals

Under certain circumstances, many molecular substances, including proteins, solidify to form crystals. In entering the crystalline state from solution, individual molecules of the substance adopt one or only a few orientations. The resulting crystal is an orderly three-dimensional array of molecules, held together by noncovalent interactions. Figure 2.3 shows such a crystalline array of molecules.

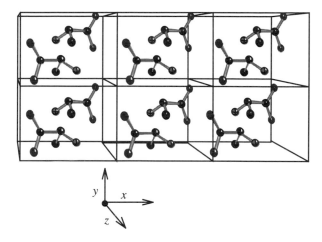

Figure 2.3 Six unit cells in a crystalline lattice. Each unit cell contains two molecules of alanine (hydrogen atoms not shown) in different orientations.

The lines in the figure divide the crystal into identical unit cells. The array of points at the corners or vertices of unit cells is called the *lattice*. The unit cell is the smallest and simplest volume element that is completely representative of the whole crystal. If we know the exact contents of the unit cell, we can imagine the whole crystal as an efficiently packed array of many unit cells stacked beside and on top of each other, more or less like identical boxes in a warehouse.

From crystallography, we obtain an image of the electron clouds that surround the molecules in the average unit cell in the crystal. We hope this image will allow us to locate all atoms in the unit cell. The location of an atom is usually given by a set of three-dimensional Cartesian coordinates, x, y, and z. One of the vertices (a lattice point or any other convenient point) is used as the origin of the unit cell's coordinate system, and is assigned the coordinates $x = 0$, $y = 0$, and $z = 0$, usually written (0,0,0). See Fig. 2.4.

B. Growing crystals

Crystallographers grow crystals of proteins by slow, controlled precipitation from aqueous solution under conditions that do not denature the protein. A number of substances cause proteins to precipitate. Ionic com-

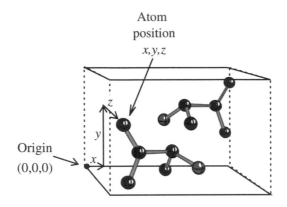

Figure 2.4 One unit cell from Fig. 2.3. The position of an atom in the unit cell can be specified by a set of spatial coordinates x,y,z.

pounds (salts) precipitate proteins by a process called "salting out." Organic solvents also cause precipitation, but they often interact with hydrophobic portions of proteins and thereby denature them. The water-soluble polymer polyethylene glycol is widely used because it is a powerful precipitant and a weak denaturant.

One simple means of causing slow precipitation is to add precipitant to an aqueous solution of protein until the precipitant concentration is just below that required to precipitate the protein. Then water is allowed to evaporate slowly, which gently raises the concentration of both protein and denaturant until precipitation occurs. Whether the protein forms crystals or instead forms a useless amorphous solid depends on many properties of the solution, including protein concentration, temperature, pH, and ionic strength. Finding the exact conditions to produce good crystals of a specific protein often requires many careful trials and is perhaps more art than science. I will examine crystallization methods in Chapter 3.

III. Collecting x-ray data

Figure 2.5 depicts, in a simple way, the collection of x-ray diffraction data. A crystal is mounted between an x-ray source and an x-ray detector. The crystal lies in the path of a narrow beam of x-rays coming from the source.

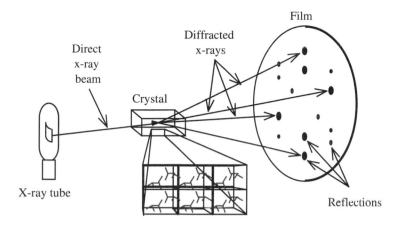

Figure 2.5 Crystallographic data collection. The crystal diffracts the source beam into many discrete beams, each of which produces a distinct spot (reflection) on the film. The positions and intensities of these reflections contain the information needed to determine molecular structures.

A simple detector is x-ray film, which when developed exhibits dark spots where x-ray beams have impinged. These spots are called *reflections* because they emerge from the crystal as if reflected from planes of atoms. Figure 2.6 shows the complex diffraction pattern of x-rays produced on film by a protein crystal. Notice that the crystal diffracts the source beam into many discrete beams, each of which produces a distinct reflection on the film. The greater the intensity of the x-ray beam that reaches a particular position, the darker the reflection.

An optical scanner precisely measures the position and the intensity of each reflection and transmits this information in digital form to a computer for analysis. The position of a reflection can be used to obtain the direction in which that particular beam was diffracted by the crystal. The intensity of a reflection is obtained by measuring the optical absorbance of the spot on the film, giving a measure of the strength of the diffracted beam that produced the spot. The computer program that reconstructs an image of the molecules in the unit cell requires these two parameters, the beam intensity and direction, for each diffracted beam.

Although film for data collection has largely been replaced by devices that feed diffraction data (positions and intensities of each reflection) directly into computers, I will continue to speak of the data as if collected on film because of the simplicity of that format, and because diffraction patterns are usually published in a form identical to their appearance on film. I will discuss other methods of collecting data in Chapter 4.

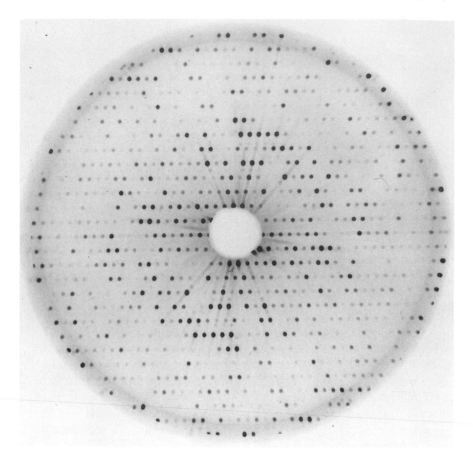

Figure 2.6 Diffraction pattern from a crystal of the MoFe (molybdenum–iron) protein of the enzyme nitrogenase from *Clostridium pasteurianum*. Notice that the reflections lie in a regular pattern, but their intensities (darkness of spots) are highly variable. Photo courtesy of Professor Jeffrey Bolin.

IV. Diffraction

A. Simple objects

You can develop some visual intuition for the information available from x-ray diffraction by examining the diffraction patterns of simple objects like spheres or arrays of spheres (Figs. 2.7–2.10). Figure 2.7 depicts dif-

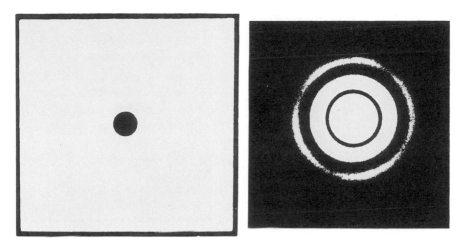

Figure 2.7 Sphere (cross section, on left) and its diffraction pattern (right).

fraction by a single sphere, shown in cross section on the left. The diffraction pattern, on the right, is a set of concentric light and dark circles.[1]

For now, just accept the fact that diffraction by a sphere produces this pattern, and think of it as the diffraction signature of a sphere. In a sense, you are already equipped to do very simple structure determination; that is, you can now recognize a simple sphere by its diffraction pattern.

B. Arrays of simple objects: Real and reciprocal lattices

Figure 2.8 depicts diffraction by a crystalline array of spheres, with a cross section of the crystal on the left, and its diffraction pattern on the right. (Spots in a diffraction pattern may be dark on a light background, as in Fig. 2.6, or light on a dark background, as in Fig. 2.8, depending on whether the pattern is collected on positive or negative film.)

The diffraction pattern, like that produced by crystalline nitrogenase (Fig. 2.6), consists of reflections (spots) in an orderly array on the film. The spacing of the reflections varies with the spacing of the spheres in their

---

[1] The patterns shown in Figs. 2.7–2.10 are actually optical diffraction patterns produced by visible laser light diffracted by arrays of holes in an opaque mask. The principles of diffraction are the same for this situation as for arrays of solid objects. These figures are from G. Harburn, C. A. Taylor, and T. R. Welberry, *Atlas of Optical Transforms,* Chapman & Hall, London, 1975 (originally published by Unwin Hyman).

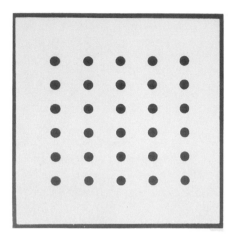 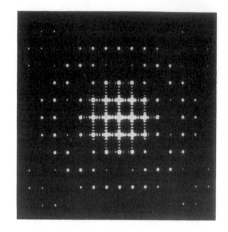

Figure 2.8 Lattice of spheres (left) and its diffraction pattern (right). If you look at the pattern and blur your eyes, you will see the diffraction pattern of a sphere.

array. Specifically, observe that although the lattice spacing of the crystal is smaller vertically, the diffraction spacing is smaller horizontally. In fact, there is a simple inverse relationship between the spacing of unit cells in the crystalline lattice, called the *real lattice*, and the spacing of reflections in the lattice on the film, which, because of its inverse relationship to the real lattice, is called the *reciprocal lattice*.

Because the real-lattice spacing is inversely proportional to the spacing of reflections, crystallographers can calculate the dimensions, in angstroms, of the unit cell of the crystalline material from the spacings of the reciprocal lattice on the x-ray film (Chapter 4). The simplicity of this relationship is a dramatic example of how the macroscopic dimensions of the diffraction pattern are connected to the submicroscopic dimensions of the crystal.

C. Intensities of reflections

Now look at the intensities of the reflections in Fig. 2.8. Some are intense ("bright"), while others are weak or perhaps missing from the otherwise evenly spaced pattern. These variations in intensity contain important information. If you blur your eyes slightly while looking at the diffraction pattern, so that you cannot see individual spots, you will see the concentric light and dark circles characteristic of diffraction by a sphere. (You just determined your first crystallographic structure.) The diffraction pattern of spheres in a lattice is simply the diffraction pattern of the average sphere in the lattice, but this pattern is incomplete. The pattern is sampled at points whose spacings vary inversely with real-lattice spacings. The pattern of varied intensities is that of the *average* sphere because all the spheres contribute to the observed pattern. To put it another way, the observed pattern of intensities is actually a superposition of the many identical diffraction patterns of all the spheres.

D. Arrays of complex objects

This relationship between (1) diffraction by a single object and (2) diffraction by many identical objects in a lattice holds true for complex objects also. Figure 2.9 depicts diffraction by six spheres that form a planar hexagon, like the six carbons in benzene.

Notice the starlike six-fold symmetry of the diffraction pattern. Again, just accept this pattern as the diffraction signature of a hexagon of spheres. (Now you can determine the structures of *two* simple objects by diffraction.) Figure 2.10 depicts diffraction by three crystalline arrays of these hexagonal objects. As before, the spacing of reflections varies reciprocally with lattice spacing, but if you blur your eyes slightly, you will see the starlike signature of a single hexagonal cluster in each diffraction pattern.

From these simple examples, you can see that the reciprocal-lattice spacing (the spacing of reflections in the diffraction pattern) is characteristic of (inversely related to) the spacing of identical objects in the crystal, while

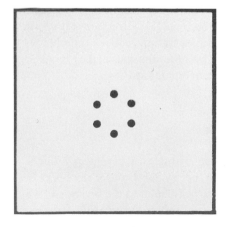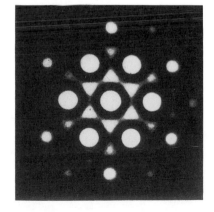

Figure 2.9 A planar hexagon of spheres (left) and its diffraction pattern (right).

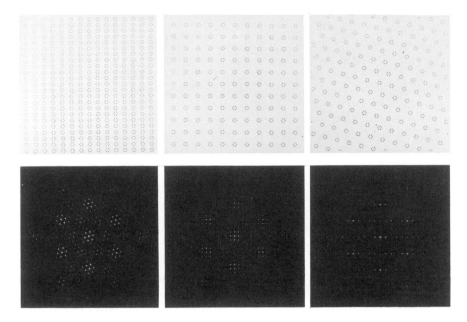

Figure 2.10 Lattices of hexagons (top) and diffraction patterns of each (bottom). If you look at each pattern and blur your eyes, you will see the diffraction pattern of a hexagon.

the reflection intensities are characteristic of the shape of the individual objects. From the reciprocal-lattice spacing in a diffraction pattern, we can compute the dimensions of the unit cell. From the intensities of the reflections, we can learn the shape of the individual molecules that compose the crystal. It is actually advantageous that the object's diffraction pattern is sampled at reciprocal-lattice positions. This sampling reduces the number of intensity measurements we must take from the film and makes it easier to program a computer to locate and measure the intensities.

E. Three-dimensional arrays

Unlike the two-dimensional arrays in these examples, a crystal is a three-dimensional array of objects. If we rotate the crystal in the x-ray beam, a different cross section of objects will lie perpendicular to the beam, and we will see a different diffraction pattern. In fact, just as the two-dimensional arrays of objects we have discussed are cross sections of objects in the three-dimensional crystal, each two-dimensional *array of reflections* (each diffraction pattern recorded on film) is a cross section of a three-dimensional lattice of reflections. Figure 2.11 shows a hypothetical three-dimensional diffraction pattern, with the reflections that would be produced by all possible orientations of a crystal in the x-ray beam.

Notice that only one plane of the three-dimensional diffraction pattern is superimposed on the film. With the crystal in the orientation shown, reflections shown in the plane of the film (solid spots) are the only reflections that produce spots on the film. In order to measure the directions and intensities of all additional reflections (shown as hollow spots), the crystallographer must collect diffraction patterns from all unique orientations of the crystal with respect to the x-ray beam. The direct result of crystallographic data collection is a list of intensities for each point in the three-dimensional reciprocal lattice. This set of data is the raw material for determining the structures of molecules in the crystal.

(*Note*: The spatial relationship involving beam, crystal, film, and reflections is more complex than shown here. I will discuss the actual relationship in Chapter 4.)

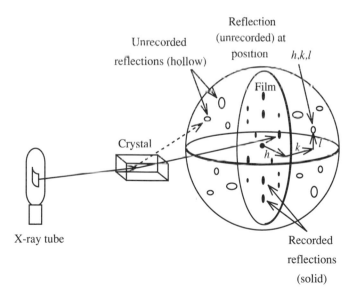

Figure 2.11 Crystallographic data collection, showing reflections measured at one particular crystal orientation (solid, on film) and those that could be measured at other orientations (hollow, within the sphere but not on the film). The relationship between measured and unmeasured reflections is more complex than shown here (see Chapter 4).

V. Coordinate systems in crystallography

Each reflection can be assigned three coordinates or indices in the imaginary three-dimensional space of the diffraction pattern. This space, the strange land where the reflections live, is called *reciprocal space*. Crystallographers usually use *h, k,* and *l* to designate individual reflections in the reciprocal space of the diffraction pattern. The central reflection (the round solid spot at the center of the film in Fig. 2.11) is taken as the origin in reciprocal space and assigned the coordinates $(h,k,l) = (0,0,0)$, usually written $hkl = 000$. (The 000 reflection is not measurable because it is always obscured by x-rays that pass straight through the crystal.) The other reflections are assigned whole-number coordinates counted from this origin, so the indices *h, k,* and *l* are integers. Thus the parameters we can measure and analyze in the x-ray diffraction pattern are the position *hkl* and the intensity I_{hkl} of each reflection. The position of a reflection is related to the angle by which the diffracted beam diverges from the source beam. For a unit cell of known dimensions, the angle of divergence uniquely specifies the indices of a reflection (see Chapter 4).

Alternatively, actual distances, rather than reflection indices, can be measured in reciprocal space. Because the dimensions of reciprocal space are the inverse of dimensions in the real space of the crystal, distances in reciprocal space are expressed in the units $Å^{-1}$ (called *reciprocal angstroms*). Roughly speaking, the inverse of the reciprocal-space distance from the origin out to the most distant measurable reflections gives the potential resolution of the model that we can obtain from the data. So a crystal that gives measurable reflections out to a distance of 1/(3 Å) from the origin should yield a model with a resolution of 3 Å.

The crystallographer works back and forth between two different coordinate systems. Let us review them briefly. The first system (see Fig. 2.4) is the unit cell (real space), where an atom's position is described by its coordinates *x,y,z*. A vertex of the unit cell, or any other convenient position, is taken as the origin, with coordinates $x,y,z = (0,0,0)$. Coordinates in real space designate real spatial positions within the unit cell. Real-space coordinates are usually given in angstroms or nanometers, or in fractions of unit cell dimensions. The second system (see Fig. 2.11) is the three-dimensional diffraction pattern (reciprocal space), where a reflection's position is described by its indices *hkl*. The central reflection is taken as the origin with the index 000 (round black dot at center of sphere). The position of a reflection is designated by counting reflections from 000, so the indices *h, k,* and *l* are integers. Distances in reciprocal space, expressed in reciprocal ang-

Figure 2.12 Fun in reciprocal space. Drawing by John O'Brien; © 1991, The New Yorker Magazine, Inc.

stroms or reciprocal nanometers, are used to judge the potential resolution that the diffraction data can yield.

Like Alice's looking-glass world, reciprocal space may seem strange to you at first (Fig. 2.12). We will see, however, that some aspects of crystallography are actually easier to understand, and some calculations are more convenient, in reciprocal space than in real space (Chapter 4).

VI. The mathematics of crystallography: A brief description

The problem of determining the structure of objects in a crystalline array from their diffraction pattern is, in essence, a matter of converting the experimentally accessible information in the reciprocal space of the diffraction pattern to otherwise inaccessible information about the real space inside the unit cell. Remember that the computer programs that make this conversion are acting as a lens to reconstruct an image from diffracted radiation. Each reflection is produced by a beam of electromagnetic radiation (x-rays), so the computations entail treating the reflections as waves and recombining these waves to produce an image of the molecules in the unit cell.

A. Wave equations: Periodic functions

Each reflection is the result of diffraction from complicated objects, the molecules in the unit cell, so the resulting wave is complicated also. Before considering how the computer represents such an intricate wave, let us consider mathematical descriptions of the simplest waves.

A simple wave, like that of visible light or x-rays, can be described by a periodic function, for instance, an equation of the form

$$f(x) = F \cos 2\pi (hx + \alpha) \tag{2.1}$$

or

$$f(x) = F \sin 2\pi (hx + \alpha) \tag{2.2}$$

In these functions, $f(x)$ specifies the vertical height of the wave at any horizontal position x along the wave. The variable x and the constant α are angles expressed in fractions of the wavelength; that is, $x = 1$ implies a position of one full wavelength (2π radians or 360°) from the origin. The constant F specifies the amplitude (the height of the crests and troughs) of the wave. For example, the crests of the wave $f(x) = 3 \cos 2\pi x$ are three times as high and the troughs are three times as deep as those of the wave $f(x) = \cos 2\pi x$ (compare b with a in Fig. 2.13).

The constant h in a simple wave equation specifies the frequency or wavelength of the wave. For example, the wave $f(x) = \cos 2\pi(5x)$ has five times the frequency (or one-fifth the wavelength) of the wave $f(x) = \cos 2\pi x$. (Compare c with a in Fig. 2.13). (In the wave equations used in this book, h takes on integral values only.)

Finally, the constant α specifies the phase of the wave, that is, the position of the wave with respect to the origin of the coordinate system on which the wave is plotted. For example, the position of the wave $f(x) = \cos 2\pi(x + \frac{1}{4})$ is shifted by $\frac{1}{4}$ of 2π radians (or one-fourth of a wavelength, or 90°) from the position of the wave $f(x) = \cos 2\pi x$ (compare d with a in Fig. 2.13). Because the wave is repetitive, with a repeat distance of one wavelength or 2π radians, a phase of $\frac{1}{4}$ is the same as a phase of $1\frac{1}{4}$, or $2\frac{1}{4}$, or $3\frac{1}{4}$, and so on. In radians, a phase of 0 is the same as a phase of 2π, or 4π, or 6π, and so on.

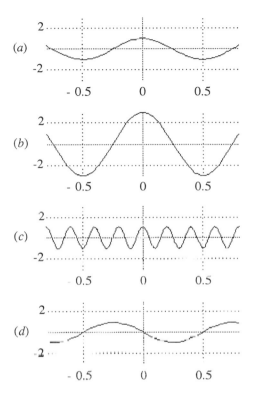

Figure 2.13 Graphs of four simple wave equations $f(x) = F \cos 2\pi(hx + \alpha)$. (a) $F = 1$, $h = 1$, $\alpha = 0$: $f(x) = \cos 2\pi(x)$. (b) $F = 3$, $h = 1$, $\alpha = 0$: $f(x) = 3 \cos 2\pi(x)$. Increasing F increases the amplitude of the wave. (c) $F = 1$, $h = 5$, $\alpha = 0$: $f(x) = \cos 2\pi(5x)$. Increasing h increases the frequency (or decreases the wavelength λ) of the wave. (d) $F = 1$, $h = 1$, $\alpha = \frac{1}{4}$: $f(x) = \cos 2\pi(x + \frac{1}{4})$. Changing α changes the position or phase of the wave.

These equations describe one-dimensional waves, in which a property (in this case, the height of the wave) varies is one direction. Visualizing a one-dimensional function $f(x)$ requires a two-dimensional graph, with the second dimension used to represent the numerical value of $f(x)$. For example, if $f(x)$ describes the electrical part of an electromagnetic wave, the x-axis is the direction the wave is moving, and the height of the wave at any position on the x-axis represents the momentary strength of the electrical field at a distance x from the origin. The field strength is in no real sense perpendicular to x, but it is convenient to use the perpendicular direction to show the numerical value of the field strength. In general, visualizing a function in n dimensions requires $n + 1$ dimensions.

B. Complicated periodic functions: Fourier series

As discussed in Section VI.A, any simple sine or cosine wave can be described by three constants: the amplitude F, the frequency h, and the phase α. It is less obvious that far more complicated waves can also be described with this same simplicity. The French mathematician Jean Baptiste Joseph Fourier (1768–1830) showed that even the most intricate periodic functions can be described as the sum of simple sine and cosine functions whose wavelengths are integral fractions of the wavelength of the complicated function. Such a sum is called a *Fourier series,* and each simple sine or cosine function in the sum is called a *Fourier term.*

Figure 2.14 shows a periodic function, called a "step function," and the beginning of a Fourier series that describes it.

A method called *Fourier synthesis* is used to compute the sine and cosine terms that describe a complex wave, which I will call the "target" of the synthesis. I will discuss the results of Fourier synthesis but not the method itself. In the example of Fig. 2.14, the first four terms produced by Fourier synthesis are shown individually (f_0 through f_3), and each is added sequentially to the Fourier series. Notice that the first term in the series, $f_0 = 1$, simply displaces the sums upward, so that they have only positive values like the target function. (Sine and cosine functions themselves have both positive and negative values, with average values of zero.) The second term, $f_1 = \cos 2\pi x$, has the same wavelength as the step function, and wavelengths of subsequent terms are simple fractions of that wavelength. (It is equivalent to say, and it is plain in the equations, that the frequencies h are simple multiples of the frequency of the step function.) Notice that the sum of only the first few Fourier terms merely approximates the target. If additional terms of shorter wavelength are computed and added, the fit of the approximated wave to the target improves, as shown by the sum of the first six terms. Indeed, using the tenets of Fourier theory, it can be proven that such approximations can be made as similar as desired to the target waveform, simply by including enough terms in the series.

Look again at the components of the Fourier series, functions f_0 through f_3. The low-frequency terms like f_1 approximate the gross features of the target wave. Higher-frequency terms like f_3 improve the approximation by filling in finer details, for example, making the approximation better in the sharp corners of the target function.

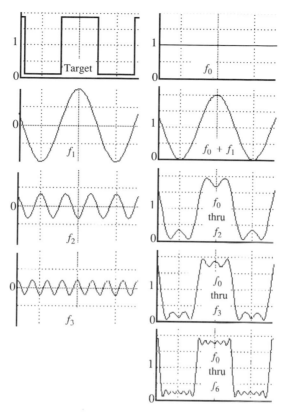

Figure 2.14 Beginning of a Fourier series to approximate a target function, in this case, a step function or square wave. $f_0 = 1$; $f_1 = \cos 2\pi(x)$; $f_2 = (-\frac{1}{3})\cos 2\pi(3x)$; $f_3 = (\frac{1}{5})\cos 2\pi(5x)$. In the left column are the target and terms f_1 through f_3. In the right column are f_0 and the succeeding sums as each term is added to f_0. Notice that the approximation improves (i.e., each successive sum looks more like the target) as the number of Fourier terms in the sum increases. In the last graph, terms f_5 and f_6 are added (but not shown separately) to show further improvement in the approximation.

C. Structure factors: Wave descriptions of x-ray reflections

Each diffracted x-ray that arrives at the film to produce a recorded reflection can also be described by a Fourier series. The Fourier series that describes a diffracted ray is called a *structure-factor equation*. The computed sum of the series for the reflection hkl is called the *structure factor* F_{hkl}. As we will

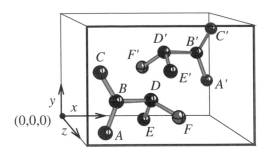

Figure 2.15 Every atom contributes to every reflection in the diffraction pattern, as described for this unit cell by Equation (2.3).

see in Chapter 4, the structure-factor equation can be written in several different ways. For example, one useful form is a series in which each Fourier term describes diffraction by one atom in the unit cell, and thus the series contains the same number of terms as the number of atoms (Fig. 2.15).

If diffraction by atom A is represented by f_A, then one diffracted ray (producing one reflection) from the unit cell of Fig. 2.15 is described by a structure-factor equation of this form:

$$F_{hkl} = f_A + f_B + \cdots + f_{A'} + f_{B'} + \cdots + f_{F'} \qquad (2.3)$$

The structure-factor equation implies, and correctly so, that each reflection on the film is the result of diffractive contributions from all atoms in the unit cell. That is, every atom in the unit cell contributes to every reflection in the diffraction pattern. The structure factor is a wave created by the superposition of many individual waves, each resulting from diffraction by an individual atom.

D. Electron-density maps

To be more precise, when we direct an x-ray beam toward a crystal, the actual diffractors of the x-rays are the clouds of electrons in the molecules of the crystal. Diffraction should therefore reveal the distribution of electrons, or the electron density, of the molecules. Electron density, of course, reflects the molecule's shape; in fact, you can think of the molecule's boundary as a van der Waals surface, the surface of a cloud of electrons that surrounds the molecule. Because, as noted earlier, protein molecules are ordered, and because, in a crystal, the molecules are in an ordered array, the electron density in a crystal can be described mathematically by a periodic function.

If we could walk through the crystal depicted in Fig. 2.3, along a linear path parallel to a cell edge, and carry with us a device for measuring electron density, our device would show us that the electron density varies along our path in a complicated periodic manner, rising as we pass through molecules, falling in the space between molecules, and repeating its variation identically as we pass through each unit cell. Because this statement is true for linear paths parallel to all three cell edges, the electron density, which describes the surface features and overall shape of all molecules in the unit cell, is a three-dimensional periodic function. I will refer to this function as $\rho(x,y,z)$, implying that it specifies a value ρ for electron density at every position x,y,z in the unit cell. A graph of the function is an image of the electron clouds that surround the molecules in the unit cell. The most readily interpretable graph is a contour map; a drawing of a surface along which there is constant electron density (refer to Plate 2a). The graph is called an *electron-density map*. The map is, in essence, a fuzzy image of the molecules in the unit cell. The goal of crystallography is to obtain the mathematical function whose graph is the desired electron-density map.

E. Electron density from structure factors

Because the electron density we seek is a complicated periodic function, it, like a structure factor, can be described as a Fourier series. Do the many structure-factor equations, each a Fourier series describing one reflection in the diffraction pattern, have any connection with the Fourier series that describes the electron density? As mentioned earlier, each structure-factor equation can be written as a Fourier series in which each term describes diffraction from one atom in the unit cell. But this is only one of many ways to write a structure-factor equation. Another way is to imagine dividing the electron density in the unit cell into many small volume elements by inserting planes parallel to the cell edges (Fig. 2.16).

These volume elements can be as small and numerous as desired. Now because the true diffractors are the clouds of electrons, each structure-factor equation can be written as a Fourier series in which each term describes diffraction by the electrons in one volume element. In this Fourier series, each term contains the average numerical value of the desired electron-density function $\rho(x,y,z)$ within one volume element. If the cell is divided into

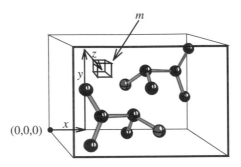

Figure 2.16 Small volume element *m* within the unit cell, one of many elements formed by subdividing the unit cell with planes parallel to the cell edges. The average electron density within *m* is ρ_m (x,y,z). Every volume element contributes to every reflection in the diffraction pattern, as described by Equation (2.4).

n elements, and the average electron density in volume element *m* is ρ_m, then one diffracted ray from the unit cell of Fig. 2.16 is described by a structure-factor equation of this form:

$$F_{hkl} = f(\rho_1) + f(\rho_2) + \cdots + f(\rho_m) + \cdots + f(\rho_n) \qquad (2.4)$$

So each reflection is described by an equation like this, giving us a large number of equations describing reflections in terms of the electron density. Is there any way to solve these equations for the function $\rho(x,y,z)$ in terms of the measured reflections? After all, structure factors like Equation (2.4) describe the reflections in terms of $\rho(x,y,z)$, which is precisely the function the crystallographer is trying to learn. I will show in Chapter 5 that a mathematical operation called the *Fourier transform* solves the structure-factor equations for the desired function $\rho(x,y,z)$, just as if they were a set of simultaneous equations describing $\rho(x,y,z)$ in terms of the amplitudes, frequencies, and phases of the reflections.

The Fourier transform describes precisely the mathematical relationship between an object and its diffraction pattern. In Figs. 2.7–2.10, the diffraction patterns are the Fourier transforms of the corresponding objects or arrays of objects. To put it another way, the Fourier transform is the lens-simulating operation that a computer performs to produce an image of molecules in the crystal. This view of $\rho(x,y,z)$ as the Fourier transform of the structure factors implies that if we can measure three parameters—amplitude, frequency, and phase—of *each* reflection, then we can obtain the function $\rho(x,y,z)$, graph the function, and "see" a fuzzy image of the molecules in the unit cell.

F. Electron density from measured reflections

Are all three of these parameters accessible in the data on our films? We will see in Chapter 5 that the measurable intensity I_{hkl} of one reflection gives the amplitude of one Fourier term in the series that describes $\rho(x,y,z)$, and that the position hkl specifies the frequency for that term. But the phase α of each reflection is not recorded on the film. In Chapter 6, we will see how to obtain the phase of each reflection, completing the information we need to calculate $\rho(x,y,z)$.

G. Obtaining a model

Having obtained $\rho(x,y,z)$, we graph the function to produce an electron-density map, an image of the molecules in the unit cell. Finally, we interpret the map by building a model that fits it (refer to Plate 2b). In interpreting the molecular image and building the model, a crystallographer takes advantage of all current knowledge about the protein under investigation, as well as knowledge about protein structure in general. Probably the most important information required is the sequence of amino acids in the protein. In a few rare instances, the amino acid sequence has been learned from the crystallographic structure. But in almost all cases, crystallographers know the sequence to start with, from the work of chemists or molecular biologists, and use it to help them interpret the image obtained from crystallography. In effect, the crystallographer starts with knowledge of the chemical structure, but without knowledge of the conformation. Interpreting the image amounts to finding a chemically realistic conformation that fits the image precisely.

A crystallographer interprets a map by displaying it on a graphics computer and building a graphics model within it. The final model must be (1) consistent with the image and (2) chemically realistic; that is, it must possess bond lengths, bond angles, conformational angles, and distances between neighboring groups that are all in keeping with established principles of molecular structure and stereochemistry. With such a model in hand, the crystallographer can begin to explore the model for clues about its function.

In Chapters 3 through 7, I will discuss in more detail the principles introduced in this chapter.

3 Protein Crystals

I. Properties of protein crystals

A. Introduction

As the term *x-ray crystallography* implies, the sample being examined is in the crystalline state. Crystals of many proteins and other biomolecules have been obtained and analyzed in the x-ray beam. A few macromolecular crystals are shown in Fig. 3.1.

In these photographs, the crystals appear much like inorganic materials such as sodium chloride. But there are several important differences between protein crystals and ionic solids.

B. Size, structural integrity, and mosaicity

Whereas inorganic crystals can often be grown to dimensions of several centimeters or larger, it is frequently impossible to grow protein crystals as large as 1 mm in their shortest dimension. In addition, larger crystals are

often twinned (two or more crystals grown into each other at different orientations) or otherwise imperfect and not usable. Roughly speaking, protein crystallography requires a crystal of at least 0.5 mm in its shortest dimension, although modern methods of data collection can sometimes succeed with smaller crystals.

Inorganic crystals derive their structural integrity from the electrostatic attraction of fully charged ions. On the other hand, protein crystals are held together by weaker forces, primarily hydrogen bonds between hydrated protein surfaces. In other words, proteins in the crystal stick to each other primarily by hydrogen bonds through intervening water molecules. Protein crystals are thus much more fragile than inorganic crystals; gentle pressure with a needle is enough to crush the hardiest protein crystal. Growing, handling, and mounting crystals for analysis thus require very gentle techniques. Protein crystals are usually harvested, examined, and mounted for crystallography within their "mother liquor," the solution in which they formed.

The textbook image of a crystal is that of a perfect array of unit cells stretching throughout. Real macroscopic crystals are actually mosaics of many submicroscopic arrays in rough alignment with each other. The result of mosaicity is that an x-ray reflection actually emerges from the crystal as a narrow cone rather than a perfectly linear beam. Thus the reflection must be measured over a very small angle, rather than at a single, well-defined angle. In protein crystals, composed as they are of relatively flexible molecules held together by weak forces, this mosaicity is more pronounced than in crystals of rigid organic or inorganic molecules, and the reflections from protein crystals suffer a greater mosaic spread than do those from more ordered crystals.

Figure 3.1 Some protein crystals grown by a variety of techniques and using a number of different precipitating agents. They are (A) deer catalase, (B) trigonal form fructose-1,6-diphosphatase from chicken liver, (C) cortisol binding protein from guinea pig sera, (D) concanavalin B from jack beans, (E) beef liver catalase, (F) an unknown protein from pineapples, (G) orthorhombic form of the elongation factor Tu from *Escherichia coli*, (H) hexagonal and cubic crystals of yeast phenylalanine tRNA, (I) monoclinic laths of the gene 5 DNA unwinding protein from bacteriophage fd, (J) chicken muscle glycerol-3-phosphate dehydrogenase, and (K) orthorhombic crystals of canavalin from jack beans. From A. McPherson, in *Methods in Enzymology*, Vol. 114, H. W. Wyckoff, C. H. W. Hirs, and S. N. Timasheff, eds., Academic Press, Orlando, Florida, 1985, p. 114. Photo and caption reprinted with permission of the author.

C. Multiple crystalline forms

In efforts to obtain crystals, or to find optimal conditions for crystal growth, crystallographers sometimes obtain a protein or other macromolecule in more than one crystalline form. Compare, for instance, Figs. 3.1A and 3.1E, which show crystals of the enzyme catalase from two different species. Although these enzymes are almost identical in molecular structure, they crystallize in different forms. In Fig. 3.1H, you can see that highly purified yeast phenylalanyl tRNA (transfer ribonucleic acid) crystallizes in two different forms. Often, the various crystal forms will differ in quality of diffraction, in ease and reproducibility of growth, and perhaps in other properties. The crystallographer must ultimately choose the best form with which to work. Quality of diffraction is the most important criterion, because it determines the ultimate quality of the crystallographic model. Among forms that diffract equally well, more symmetric forms are usually preferred because they require less data collection (see Chapter 4).

D. Water content

Early protein crystallographers, proceeding by analogy with studies of other crystalline substances, examined dried protein crystals and obtained no diffraction patterns. Thus x-ray diffraction did not appear to be a promising tool for analyzing proteins. In 1934, J. D. Bernal and Dorothy Crowfoot (later Hodgkin) measured diffraction from pepsin crystals still in the liquid from which they crystallized. Bernal and Crowfoot recorded sharp diffraction patterns, with reflections out to distances in reciprocal space that correspond in real space to the distances between atoms. The announcement of their success was, in effect, a birth announcement for protein crystallography.

Careful analysis of electron-density maps usually reveals many ordered water molecules on the surface of crystalline proteins (Plate 3). Additional disordered water is presumed to occupy regions of low density between the ordered particles. The quantity of water varies among proteins and even among different crystal forms of the same protein. The number of detectable ordered water molecules averages about one per amino-acid residue in the protein. Both ordered and disordered water molecules are essential to crystal integrity, and drying destroys the crystal structure. For this reason, protein crystals are subjected to x-ray analysis in a very humid atmosphere or in a solution that will not dissolve them, such as the mother liquor.

NMR analysis of protein structure suggests that the ordered water molecules seen by x-ray diffraction on protein surfaces have very short residence times in solution. Thus these molecules may be of little use to an understanding of protein function. However, ordered water is of great importance to the crystallographer. As the structure determination progresses, ordered water becomes visible in the electron-density map. Assignment of water molecules to these isolated areas of electron density improves the overall accuracy of the model, and, for reasons I will discuss in Chapter 7, improvements in accuracy in one area of the model give accompanying improvements in other regions.

II. Evidence that solution and crystal structures are similar

Knowing that crystallographers study proteins in the crystalline state, you may be wondering if these molecules are altered when they crystallize, and whether the structure revealed by x-rays is pertinent to molecular action in solution. Crystallographers worry about this problem also, and with a few proteins it has been found that crystal structures are in conflict with chemical or spectroscopic evidence about the protein in solution. These cases are rare, however, and the large majority of crystal structures appear to be identical to the solution structure. Because of the slight possibility that crystallization will alter molecular structure, an essential part of any structure determination project is an effort to show that the crystallized protein is not significantly altered.

A. Proteins retain their function in the crystal

Probably the most convincing evidence that crystalline structures can safely be used to draw conclusions about molecular function is the observation that many macromolecules are still functional in the crystalline state. For example, substrates added to suspensions of crystalline enzymes are converted to product, albeit at reduced rates, suggesting that the enzymes' catalytic and binding sites are intact. The lower rates of catalysis can be accounted for by the reduced accessibility of active sites within the crystal, in comparison to solution.

In a dramatic demonstration of the persistence of protein function in the crystalline state, crystals of deoxyhemoglobin shatter in the presence of oxygen. Hemoglobin molecules are known to undergo a substantial conformational change when they bind oxygen. The conformation of oxyhemoglobin is apparently incompatible with the constraints on deoxyhemoglobin in crystalline form, and so oxygenation disrupts the crystal.

It makes sense, therefore, after obtaining crystals of a protein and before embarking on the strenuous process of obtaining a structure, to determine whether the protein retains its function in the crystalline state. If the crystalline form is functional, the crystallographer can be confident that the model will show the molecule in its functional form.

B. X-ray structures are compatible with other structural evidence

Further evidence for the similarity of solution and crystal structures is the compatibility of crystallographic models with the results of chemical studies on proteins. For instance, two reactive groups in a protein might be linked by a cross-linking reagent, demonstrating their nearness. In all cases, the groups shown to be near each other by such studies have been found near each other in the crystallographic model.

In a few recent cases, both NMR and x-ray methods have been used to determine the structure of the same molecule. Plate 4 shows the α-carbon backbones of two models of the protein thioredoxin from the bacterium *Escherichia coli*. The green model was obtained by x-ray crystallography and the white model by NMR. Clearly the two methods produce similar models. This and other NMR-derived models confirm that protein molecules are very similar in crystals and in solution. In some cases, small differences are seen and can usually be attributed to crystal packing. Often these packing effects are detectable in the crystallographic model itself. For instance, in the crystallographic model of cytoplasmic malate dehydrogenase, whose functional form is a dimer, an external loop has different conformations in the two molecules of one dimer. On examination of the dimer in the context of neighboring dimers, it can be seen that one molecule of each pair lies very close to a molecule of a neighboring pair. It was thus inferred that the observed difference between the oligomers in a dimer is due to crystal packing and, further, that the unaffected molecule of each pair is probably more like the enzyme in solution.

C. Other evidence

In a few cases, the structure of a protein has been obtained from more than one type of crystal. The resulting models were identical, suggesting that the molecular structure was not altered by crystallization.

Recall that stable protein crystals contain a large amount of both ordered and disordered water molecules. As a result, the proteins in the crystal are still in the aqueous state, subject to the same solvent effects that stabilize the structure in solution. Thus, it is less surprising that proteins retain their solution structure in the crystal.

III. Growing protein crystals

A. Introduction

Crystals suffer damage in the x-ray beam, due primarily to free radicals generated by x-rays. For this reason, a full structure determination project usually consumes many crystals. I will now consider the problem of developing a reliable, reproducible source of protein crystals. This entails not only growing good crystals of the pure protein but also obtaining derivatives, or crystals of the protein in complex with various ligands. For example, in addition to pursuing the structures of proteins themselves, crystallographers also seek structures of proteins in complexes with ligands such as cofactors, substrate analogs, inhibitors, and allosteric effectors. Structure determination then reveals the details of protein–ligand interactions, giving insight into protein function.

Another vital type of ligand is a heavy-metal atom or ion. Crystals of protein/heavy-metal complexes, often called *heavy-atom derivatives*, are usually needed in order to solve the phase problem mentioned in Chapter 2 (Section VII.F). I will show in Chapter 6 that, for the purpose of obtaining phases, it is crucial that heavy-atom derivatives possess the same unit-cell dimensions and symmetry, and the same protein conformation, as crystals of the pure protein, which in discussions of derivatives are called "native" crystals. So in most structure projects, the crystallographer must produce both native and derivative crystals under the same or very similar circumstances.

B. Growing crystals: Basic procedure

Crystals of an inorganic substance can often be grown by making a hot, saturated solution of the substance and then slowly cooling it. Polar organic compounds can sometimes be crystallized by similar procedures, or by slow precipitation from aqueous solutions by addition of organic solvents. If you work with proteins, just the mention of these conditions probably makes you cringe. Proteins, of course, are usually denatured by heating or exposure to organic solvents, so techniques used for small molecules are not appropriate. In the most common methods of growing protein crystals, purified protein is dissolved in an aqueous buffer containing a precipitant, such as ammonium sulfate or polyethylene glycol, at a concentration just below that necessary to precipitate the protein. Then water is removed by controlled evaporation to produce precipitating conditions, which are maintained until crystal growth ceases.

One widely used technique is vapor diffusion, in which the protein/precipitant solution is allowed to equilibrate in a closed container with a larger aqueous reservoir whose precipitant concentration is optimal for producing crystals. An example of this technique is the "hanging-drop method" (Fig. 3.2).

Less than 25 μL of the solution of purified protein is mixed with an equal amount of the reservoir solution, giving a precipitant concentration about 50% of that required for protein crystallization. This solution is suspended as a droplet underneath a cover slip, which is sealed onto the top of the reservoir with a stopcock or vacuum grease. Because the precipitant is the major solute present, vapor diffusion in this closed system results in net trans-

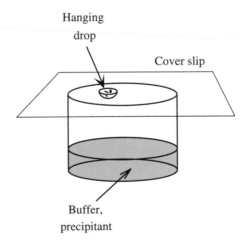

Figure 3.2 Growing crystals by the hanging-drop method. The droplet hanging under the cover slip contains buffer, precipitant, protein, and, if all goes well, growing protein crystals.

fer of water from the protein solution to the reservoir, until the precipitant concentration is the same in both solutions. Because the reservoir is much larger than the protein solution, the final concentration of the precipitant in the protein solution is nearly equal to that in the reservoir. When the system comes to equilibrium, net transfer of water ceases, and the protein solution is maintained at the optimal precipitant concentration. In this way, the precipitant concentration in the protein solution rises to the optimal level for crystallization and remains there without overshooting, because at equilibrium, the vapor pressure in the closed system equals the inherent vapor pressure of both protein solution and reservoir.

Frequently the crystallographer obtains many small crystals instead of a few that are large enough for diffraction measurements. If many crystals grow at once, the supply of dissolved protein will be depleted before crystals are large enough to be useful. Small crystals of good quality can be used as seeds to grow larger crystals. The experimental setup is the same as before, except that each hanging droplet is seeded with a few small crystals. Crystals may grow from seeds up to 10 times faster than they grow anew, so most of the dissolved protein goes into only a few crystals.

C. Growing derivative crystals

Crystallographers obtain the derivatives needed for phase determination and for studying protein–ligand interactions by two methods, cocrystallizing protein and ligand, and soaking preformed protein crystals in motherliquor solutions containing ligand.

It is sometimes possible to obtain crystals of protein–ligand complexes by crystallizing protein and ligand together, a process called *cocrystallization*. For example, a number of NAD^+ dependent dehydrogenases readily crystallize as NAD^+ or NADH complexes from solutions containing these cofactors. Cocrystallization is the only method for producing crystals of proteins in complexes with large ligands, such as nucleic acids or other proteins.

A second means of obtaining crystals of protein–ligand complexes is to soak protein crystals in mother liquor that contains ligand. As mentioned earlier, proteins retain their binding and catalytic functions in the crystalline state, and ligands can diffuse to active sites and binding sites through channels of water in the crystal. Soaking is usually preferred over cocrystallization when the crystallographer plans to compare the structure of a pure protein with that of a protein–ligand complex. Soaking preformed protein crystals with ligands is more likely to produce crystals of the same form and unit-cell dimensions as those of pure protein, so this method is recommended for first attempts to make heavy-atom derivatives.

D. Finding optimal conditions for crystal growth

Many variables influence the formation of macromolecular crystals. These include obvious ones like protein purity, concentrations of protein and precipitant, pH, and temperature, as well as more subtle ones like cleanliness, vibration and sound, convection, source and age of the protein, and the presence of ligands. Clearly, the problem of developing a reliable source of crystals entails controlling and testing a large number of parameters. (The difficulty and importance of obtaining good crystals has even prompted the invention of crystallization robots that can be programmed to set up many trials under systematically varied conditions.)

The complexity of this problem is illustrated in Fig. 3.3, which shows the effects of varying just two parameters, the concentrations of protein (in this case, the enzyme lysozyme) and precipitant (NaCl). Notice the effect of slight changes in concentration of either protein or precipitant on the rate of crystallization, as well as the size and quality of the resulting crystals.

A sample scheme for finding optimum crystallization conditions is to determine the effect of pH on precipitation with a given precipitant, repeat this determination at various temperatures, and then repeat these experiments with different precipitating agents. For such surveys of crystallization conditions, multiple batches of crystals can be grown conveniently by the hanging-drop method in clear plastic tissue-culture trays of 24 or more wells, each with its own cover slip. This apparatus has the advantage that the growing crystals can be observed through the cover slips with a dissecting microscope. Then, once the ideal conditions are found, many small batches of crystals can be grown at once, and each batch can be harvested without disturbing the others.

When varying the more conventional parameters fails to produce good crystals, the crystallographer may take more drastic measures. Sometimes limited digestion of the protein by a proteolytic enzyme removes a disordered surface loop, resulting in a more rigid, hydrophilic, or compact molecule that forms better crystals. A related measure is adding a ligand, such as a cofactor, that is known to bind tightly to the protein. The protein/cofactor complex may be more likely to crystallize than the free protein, either because the complex is more rigid than the free protein or because the cofactor induces a conformational change that makes the protein more amenable to crystallizing.

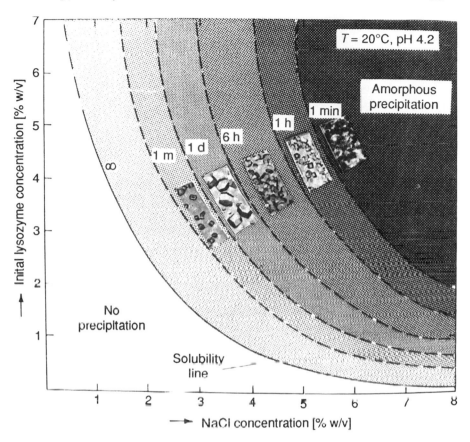

Figure 3.3 Schematic map of crystallization kinetics as a function of lysozyme and NaCl concentration obtained from a matrix of dishes. Inserts show photographs of dishes obtained 1 month after preparation of solutions. From G. Feher and Z. Kam, in *Methods in Enzymology*, Vol. 114, H. W. Wyckoff, C. H. W. Hirs, and S. N. Timasheff, eds., Academic Press, Orlando, Florida, 1985, p. 90. Photo and caption reprinted with permission of the author.

Many membrane-associated proteins will not dissolve in aqueous buffers and tend to form amorphous precipitates instead of crystals. The intractability of such proteins often results from hydrophobic domains or surface regions that are normally associated with the interior of membranes. In a few cases, such proteins have been crystallized in the presence of detergents, which coat the hydrophobic portion, decorating it with the ionic groups of detergent, and thus rendering it more soluble in water. Also, limited proteolysis of membrane-associated proteins might remove an exposed hydrophobic portion, leaving a crystallizable fragment that is more like a typical water-soluble protein.

When these drastic measures are required to yield good crystals, the crystallographer is faced with the question of whether the resulting fragment is worthy of the arduous effort to determine its structure. This question is similar to the basic issue of whether a protein has the same structure in crystal and in solution, and the question must be answered in the same way. Specifically, it may be possible to demonstrate that the fragment maintains at least part of the biological function of the intact molecule and, further, that this function is retained after crystallization.

IV. Judging crystal quality

The acid test of a crystal's suitability for structure determination is, of course, its capacity to give sharp diffraction patterns with clear reflections at large angles from the x-ray beam. A brief inspection of crystals under a low-power light microscope can also provide some insight into quality and can help the crystallographer pick out the most promising crystals.

Desirable visible characteristics of crystals include optical clarity, smooth faces, and sharp edges. Broken or twinned crystals sometimes exhibit dark cleavage planes within an otherwise clear interior. Depending on the lattice type (Chapter 4) and the direction of viewing relative to unit-cell axes, some crystals strongly rotate plane-polarized light. This property is easily observed by examining the crystal between two polarizers, one fixed and one rotatable, under a microscope. On rotation of the movable polarizer, a good-quality crystal will usually brighten and darken sharply.

Another useful physical property of the crystal is its density, which can be used to determine several useful microscopic properties, including the protein molecular weight, the protein/water ratio in the crystal, and the number of protein molecules in each asymmetric unit (defined below). Molecular weights from crystal density are more accurate than those from electrophoresis or most other methods (except mass spectrometry) and are not affected by dissociation or aggregation of protein molecules. The protein/water ratio is used to clarify electron-density maps prior to interpretation (Chapter 7). If the unit cell is symmetric (Chapter 4), it can be subdivided into two or more equivalent parts called *asymmetric units*. For interpreting electron-density maps, it is helpful to know the number of protein molecules per asymmetric unit.

Crystal density is measured in a graduated cylinder by suspending the crystal in a density gradient made by mixing water-saturated organic solvents such as xylene and carbon tetrachloride. The crystal will settle through the liquid until its density matches that of the liquid mixture and then remain suspended there. Drops of salt solutions of known density are used to calibrate the gradient.

The product of the crystal density and the unit-cell volume (determined from crystallographic analysis, Chapter 4) gives the total mass within the unit cell. This quantity, expressed in daltons, is the sum of all atomic masses in one unit cell. If the protein molecular mass and the number of protein molecules per unit cell are known, the remainder of the cell can be assumed to be water, thus establishing the protein/water ratio.

It can be shown that the molecular weight of protein in each asymmetric unit is given by

$$M_p = \frac{N \cdot V (D_c - D_w)}{n (1 - v_p D_w)} \tag{3.1}$$

in which D_c and D_w are densities of crystal and water, N is Avogadro's number, V is the volume of the unit cell, v_p is the partial specific volume of the protein, and n is the number of protein molecules of molecular mass M_p in each unit cell. The partial specific volume of the protein can be determined from its amino-acid composition (percent of each amino acid) by simply averaging the partial specific volumes of the component amino acids (obtained from tables). Thus, if the protein molecular weight is known, n can be computed. Because n is an integer, it can be determined from even a rough molecular weight, taking the integer nearest the computed result. Then substitution of the correct integral value of n into Equation (3.1) gives a precise value of M_p.

Once the crystallographer has a reliable source of suitable crystals, data collection can begin.

4 Collecting Diffraction Data

I. Introduction

In this chapter, I will discuss the geometric principles of diffraction, revealing, in both the real space of the crystal's interior and in reciprocal space, the conditions that produce reflections. I will show how these conditions allow the crystallographer to determine the dimensions of the unit cell and the symmetry of its contents, and how these factors determine the strategy of data collection. Finally, I will look at the devices used to produce and detect x-rays and to measure precisely the intensities and positions of reflections.

II. Geometric principles of diffraction

W. L. Bragg showed that the angles at which diffracted beams emerge from a crystal can be computed by treating diffraction as if it were reflection from sets of equivalent, parallel planes of atoms in a crystal. (This is why each spot in the diffraction pattern is called a *reflection*.) I will first describe how crystallographers denote the planes that contribute to the diffraction pattern.

A. The generalized unit cell

The dimensions of a unit cell are designated by six numbers: the lengths of three unique edges **a**, **b**, and **c**; and three unique angles α, β, and γ (Fig. 4.1). [Notice the use of bold type in naming the unit-cell edges or the axes that correspond to them. I will use bold letters (**a**, **b**, **c**) to signify the edges or axes themselves, and letters in italics (*a*, *b*, *c*) to specify their length. Thus *a* is the length of unit-cell edge **a**, and so forth.]

A cell in which $a \neq b \neq c$ and $\alpha \neq \beta \neq \gamma$, as in Fig. 4.1, is called *triclinic*. If $a \neq b \neq c$, $\alpha = \gamma = 90°$, and $\beta > 90°$, the cell is *monoclinic*. If $a = b = c$, $\alpha = \beta = 90°$, and $\gamma = 120°$, the cell is *hexagonal*. For cells in which all three cell angles are 90°, if $a = b = c$, the cell is *cubic*; if $a = b \neq c$, the cell is *tetragonal*; and if $a \neq b \neq c$, the cell is *orthorhombic*. The most convenient coordinate systems for crystallography adopt coordinate axes based on the directions of unit-cell edges. For cells in which at least one cell angle is not 90°, the coordinate axes are not the familiar orthogonal (mutually perpendicular) *x*, *y*, and *z*. In this book, for clarity, I will consider only unit cells and coordinate systems with orthogonal axes ($\alpha = \beta = \gamma = 90°$), and I will use orthorhombic systems most often, making it easy to distinguish the three cell edges. In such systems, the **a** edges of the cell are parallel to the *x*-axis of an orthogonal coordinate system, edges **b** are parallel to *y*, and edges **c** are parallel to *z*. Bear in mind, however, that the principles discussed here can be generalized to all unit cells.

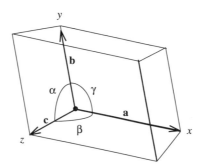

Figure 4.1 General (triclinic) unit cell, with edges **a**, **b**, and **c**, and angles α, β, and γ.

B. Indices of the atomic planes in a crystal

The most obvious sets of planes in a crystalline lattice are those determined by the faces of the unit cells. These and all other regularly spaced planes that can be drawn through lattice points can be thought of as sources of diffraction, and can be designated by a set of three numbers called *lattice indices*. Three indices *hkl* identify a particular set of equivalent, parallel planes. The index *h* gives the number of planes in the set per unit cell in the *x* direction, or equivalently, the number of parts into which the set of planes cut the **a** edge of each cell. The indices *k* and *l* specify how many such planes exist per unit cell in the *y* and *z* directions. An equivalent way to determine the indices of a set of planes is to start at any lattice point and move out into the unit cell away from the plane cutting that lattice point. If the first plane encountered cuts the **a** edge at some fraction $1/n_a$ of its length, and the same plane cuts the **b** edge at some fraction $1/n_b$ of its length, then the *h* index is n_a and the *k* index is n_b (examples given below). Indices are written in parentheses when referring to the set of planes; hence, the planes having indices *hkl* are the (*hkl*) planes.

In Fig. 4.2, each face of an orthorhombic unit cell is labeled with the indices of the set of planes that includes that face. (The crossed arrows lie on the labeled face.)

The set of planes including and parallel to the **bc** face, and hence normal to the *x*-axis, is designated (100), because there is one such plane per lattice point in the *x* direction. In like manner, the planes parallel to and including the **ac** face are called (010) planes (one plane per lattice point along *y*). Finally, the **ab** faces of the cell determine the (001) planes. In the Bragg model of diffraction as reflection from parallel sets of planes, any of these sets of planes can be the source of one diffracted x-ray beam. (Remember that an entire set of parallel planes, not just one plane, acts as a single diffractor and produces one reflection.) But if these three sets of planes were

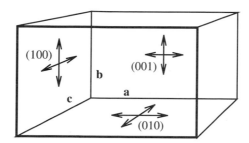

Figure 4.2 Indices of faces in an orthorhombic unit cell.

the only diffractors, the number of diffracted beams would be small, and the information obtainable from diffraction would be very limited.

In Fig. 4.3, an additional set of planes, and thus an additional source of diffraction, is indicated. The lattice (dark lines) is shown in section parallel to the **ab** faces or the *xy* plane. The dashed lines represent the intersection of a set of equivalent, parallel planes that are perpendicular to the *xy* plane of the paper. Note that the planes cut each **a** edge into two parts and each **b** edge into one part, so these planes have indices 210. Because all (210) planes are parallel to the *z*-axis (which is perpendicular to the plane of the paper), the *l* index is zero. [Or equivalently, because the planes are infinite in extent, and are coincident with **c** edges, and thus do not cut edges parallel to the *z*-axis, there are zero (21$\underline{0}$) planes per unit cell in the *z* direction.] As another example, for any plane in the set shown in Fig. 4.4, the first plane encountered from any lattice point cuts that unit cell at *a*/2 and *b*/3, so the indices are 230.

All planes perpendicular to the *xy* plane have indices *hk*0. Planes perpendicular to the *xz* plane have indices *h*0*k*, and so forth. Many additional sets of planes are not perpendicular to *x*, *y*, or *z*. For example, the (234) planes cut the unit cell edges **a** into two parts, **b** into three parts, and **c** into four parts. See Fig. 4.5.

Finally, indices can be negative as well as positive. The (210) planes are the same as (−2 −1 0), while the (2 −1 0) or (−2 1 0) planes tilt in the direction

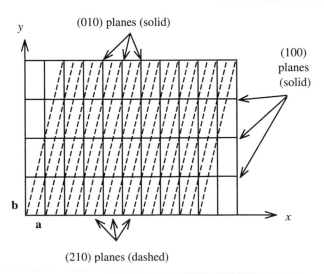

Figure 4.3 (210) planes in a two-dimensional section of lattice.

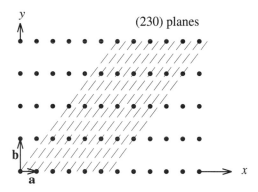

Figure 4.4 (230) planes in a two-dimensional section of lattice.

opposite to the (210) planes (Fig. 4.6). (The negative signs are often printed on top of the indices, but for clarity I will present them as shown here.)

In Bragg's way of looking at diffraction as reflection from sets of planes in the crystal, each set of parallel planes described here (as well as each additional set of planes interleaved between these sets) is treated as an independent diffractor and produces a single reflection. This model is useful for determining the geometry of data collection. Later, when I discuss structure determination, I will consider another model, in which each atom or each small volume element of electron density is treated as an independent diffractor, represented by one term in a Fourier series that describes each reflection. Bragg's model tells us where to look for the data. The Fourier series model tells us what the data have to say about molecular structure.

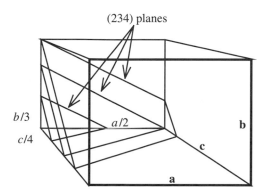

Figure 4.5 The intersection of three (234) planes with a unit cell.

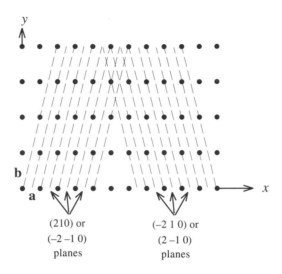

Figure 4.6 The (210) and (−2 −1 0) planes are identical. The (210) planes tilt in the opposite direction from (−1 2 0) and (−2 1 0) planes.

C. Conditions that produce diffraction: Bragg's law

Notice that the different sets of equivalent parallel planes in the preceding figures have different interplanar spacing *d*. Among sets of planes (*hkl*), interplanar spacing decreases as any index increases. Bragg showed that a set of parallel planes with index *hkl* and interplanar spacing d_{hkl} produces a diffracted beam when x-rays of wavelength λ impinge on the planes at an angle θ and are reflected at the same angle, only if θ meets the condition

$$2d_{hkl}\sin\theta = n\lambda \tag{4.1}$$

where *n* is an integer. The geometric construction in Fig. 4.7 demonstrates the conditions necessary for producing a strong diffracted ray. The dots represent two parallel planes of lattice points with interplanar spacing d_{hkl}. Two rays R_1 and R_2 are reflected from them at angle θ.

Lines *AC* are drawn from the point of reflection *A* of R_1 perpendicular to the ray R_2. If ray R_2 is reflected at *B*, then the diagram shows that R_2 travels the same distance as R_1 plus an added distance 2*BC*. Because *AB* in the small triangle *ABC* is perpendicular to the atomic plane, and *AC* is perpendicular to the incident ray, the angle *CAB* equals θ, the angle of incidence. (Two angles are equal if corresponding sides are perpendicular.) Since *ABC*

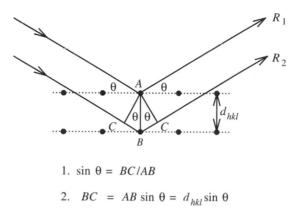

1. $\sin \theta = BC/AB$

2. $BC = AB \sin \theta = d_{hkl} \sin \theta$

Figure 4.7 Conditions that produce strong diffracted rays. If the additional distance traveled by the more deeply penetrating ray R_2 is an integral multiple of λ, then rays R_1 and R_2 interfere constructively.

is a right triangle, the sine of angle θ is BC/AB or BC/d_{hkl}. Thus BC equals $d_{hkl} \sin \theta$, and the additional distance $2BC$ traveled by ray R_2 is $2d_{hkl} \sin \theta$.

If this difference in path length for rays reflected from successive planes is equal to an integral number of wavelengths of the impinging x-rays (that is, if $2d_{hkl} \sin \theta = n\lambda$), then the rays reflected from successive planes emerge from the crystal in phase with each other, interfering constructively to produce a strong diffracted beam. For other angles of incidence θ' (where $2d_{hkl} \sin \theta'$ *does not* equal an integral multiple of λ), waves emerging from successive planes are out of phase, so they interfere destructively and no beam emerges at that angle. Think of it this way: If x-rays impinge at an angle θ' that does not satisfy the Bragg conditions, then for every reflecting plane p, there will exist, at some depth in the crystal, another parallel plane p' producing a wave precisely ($180°$) out of phase with that from p, and thus precisely canceling the wave from p. So all such waves will be canceled by destructive interference, and no diffracted ray will emerge at the angle θ'. Strong diffracted rays emerge from (hkl) planes of spacing d_{hkl} only at angles θ for which $2d_{hkl} \sin \theta = n\lambda$.

Notice that the angle of diffraction θ is inversely related to the interplanar spacing d_{hkl} ($\sin \theta$ is proportional to $1/d_{hkl}$). This implies that large unit cells, with large spacings, give small angles of diffraction and hence produce many reflections that fall within a convenient angle from the incident beam. On the other hand, small unit cells give large angles of diffraction, producing fewer measurable reflections. In a sense, the number of measurable reflections depends on how much information is present in the unit cell. Large cells contain many atoms and thus more information, and they produce more information in the diffraction pattern. Small unit cells contain fewer atoms, and diffraction from them contains less information.

It is not coincidental that I use the variable names h, k, and l for both the indices of planes in the crystal and the indices of reflections in the diffraction pattern (Chapter 2, Section V). I will show below that in fact the set of planes (hkl) produces the reflection hkl of the diffraction pattern. In the terms used in Chapter 2, each set of parallel planes in the crystal produces one reflection, or one term in the Fourier series that describes the electron density within the unit cell. The intensity of that reflection depends on the electron distribution and density along the planes that produce the reflection.

D. The reciprocal lattice

Now let us consider the Bragg conditions from another point of view, in reciprocal space. Before looking at diffraction from this vantage point, I will define and tell how to construct a new lattice, the reciprocal lattice, in what will at first seem an arbitrary manner. But I will then show that the points in this reciprocal lattice are guides that tell the crystallographer the angles at which strong reflections will occur.

Figure 4.8a shows an **ab** section of lattice with an arbitrary lattice point O chosen as the origin of the reciprocal lattice I am about to define. This point is thus the origin for both the real and reciprocal lattices. Each plus symbol (+) in the figure is a real-lattice point.

Through a neighboring lattice point N, draw one plane from each of the sets (110), (120), (130), and so forth. From the origin, draw a line normal to the (110) plane. Make the length of this line $1/d_{110}$, the inverse of the interplanar spacing d_{110}. Define the reciprocal-lattice point 110 as the point at the end of this line (heavy dot). Now repeat the procedure for the (120) plane, drawing a line from O normal to the (120) plane, and of length $1/d_{120}$. Because d_{120} is smaller than d_{110} (recall that d decreases as indices increase), this second line is longer than the first. The end of this line defines a second reciprocal-lattice point, with indices 120 (heavy dot). Repeat for the planes (130), (140), and so forth.

Now continue this operation for planes (210), (310), (410), and so on, defining reciprocal-lattice points 210, 310, 410, and so on (Fig. 4.8b). Note that the points defined by this operation form a lattice, with the arbitrarily chosen real-lattice point as the origin (indices 000). This new lattice is the reciprocal lattice. The planes $hk0$, $h0k$, and $0kl$ correspond, respectively, to the xy, xz, and yz planes. They intersect at the origin and are called the *zero-level planes* in this lattice. Other planes of reciprocal-lattice points parallel to the zero-level planes are called *upper-level planes*.

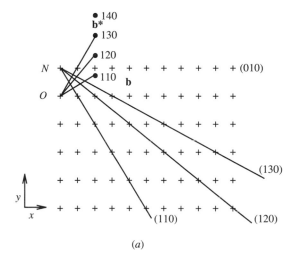

(a)

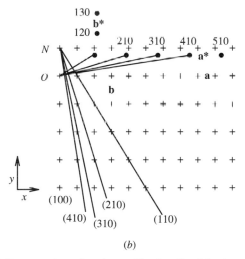

(b)

Figure 4.8 (*a*) Construction of reciprocal lattice. Real-lattice points are plus signs (+), and reciprocal-lattice points are dots. Notice the real cell edges **b** and the reciprocal cell edges **b***. (*b*) Continuation of (*a*). Notice the real cell edges **a** and the reciprocal cell edges **a***.

We can also speak of the reciprocal unit cell in such a lattice (Fig. 4.9). If the real unit-cell angles α, β, and γ are 90°, the reciprocal unit cell has axes **a*** lying along real unit-cell edge **a**, **b*** lying along **b**, and **c*** along **c**. The lengths of edges **a***, **b***, and **c*** are reciprocals of the lengths of corresponding real cell edges **a**, **b**, and **c**: $a^* = 1/a$, and so forth. If axial lengths

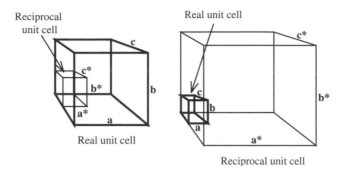

Figure 4.9 Reciprocal unit cells of large and small real cells.

are expressed in angstroms, then reciprocal-lattice spacings are in the unit $1/\text{Å}$ or Å^{-1} (reciprocal angstroms). For real unit cells with nonorthogonal axes, the spatial relationships between the real and reciprocal unit-cell edges are more complicated, and I will not make use of them in this book.

Now envision this lattice of imaginary points in the same space occupied by the crystal. For a small real unit cell, interplanar spacings d_{hkl} are small, and hence the lines from the origin to the reciprocal-lattice points are long. Therefore, the reciprocal unit cell is large, and lattice points are widely spaced. On the other hand, if the real unit cell is large, the reciprocal unit cell is small and reciprocal space is densely populated with reciprocal-lattice points.

The reciprocal lattice is spatially linked to the crystal because of the way the lattice points are defined, so if we rotate the crystal, the reciprocal lattice rotates with it. So now when you think of a crystal, and imagine the many identical unit cells stretching out in all directions (real space), imagine also a lattice of points in reciprocal space, points whose lattice spacing is inversely proportional to the interplanar spacings within the crystal.

E. Bragg's law in reciprocal space

Now I will look at diffraction from within reciprocal space. I will show that the reciprocal-lattice points give the crystallographer a convenient way to compute the direction of diffracted beams from all sets of parallel planes in the crystalline lattice (real space). This demonstration entails showing how each reciprocal-lattice point must be arranged with respect to the x-ray beam in order to satisfy Bragg's law and produce a reflection from the crystal.

Figure 4.10*a* shows an **a*b*** plane of a reciprocal lattice. Assume that an x-ray beam (arrow *XO*) impinges on the crystal along this plane.

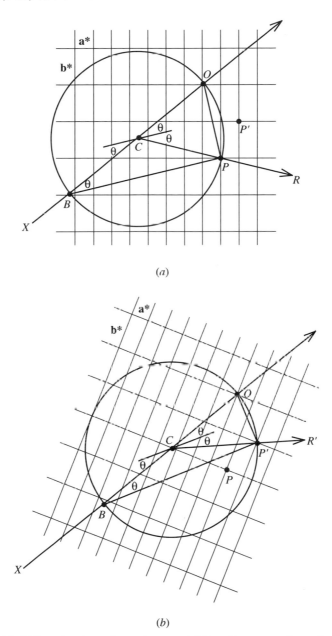

(a)

(b)

Figure 4.10 Diffraction in reciprocal space. (*a*) Ray *R* emerges from the crystal when reciprocal-lattice point *P* intersects the circle. (*b*) As the crystal rotates about point *O*, point *P'* intersects the circle, producing ray *R'*.

Point O is arbitrarily chosen as the origin of the reciprocal lattice. (Remember that O is also a real-lattice point in the crystal.) We imagine the x-ray beam passing through O along the line XO (arrow). Draw a circle of radius $1/\lambda$ having its center C on XO and passing through O. This circle represents the wavelength of the x-rays in reciprocal space. (If the wavelength is λ in real space, it is $1/\lambda$ in reciprocal space.) Rotating the crystal about O rotates the reciprocal lattice about O, successively bringing reciprocal-lattice points like P and P' into contact with the circle. In Fig. 4.10a, P (whose indices are hkl) is in contact with the circle, and the lines OP and BP are drawn. The angle PBO is θ. Because the triangle PBO is inscribed in a semicircle, it is a right triangle and

$$\sin\theta = \frac{OP}{BO} = \frac{OP}{2/\lambda} \tag{4.2}$$

Rearranging (4.2) gives

$$2\frac{1}{OP}\sin\theta = \lambda \tag{4.3}$$

Because P is a reciprocal-lattice point, the length of the line OP is $1/d_{hkl}$, where h, k, and l are the indices of the set of planes represented by P. (Recall from the construction of the reciprocal lattice that the length of a line from O to a reciprocal-lattice point hkl is $1/d_{hkl}$.) So $1/OP = d_{hkl}$ and

$$2d_{hkl}\sin\theta = \lambda \tag{4.4}$$

which is Bragg's law with $n = 1$.

In Fig. 4.10b, the crystal, and hence the reciprocal lattice, has been rotated until P', with indices $h'k'l'$, touches the circle. The same construction as in Fig. 4.10a now shows that

$$2d_{h'k'l'}\sin\theta = \lambda \tag{4.5}$$

We can conclude that whenever the crystal is rotated so that a reciprocal-lattice point comes in contact with this circle of radius $1/\lambda$, Bragg's law is satisfied and a reflection occurs. What direction does the reflected beam take?

Recall (from construction of the reciprocal lattice) that the line defining a reciprocal-lattice point is normal to the set of planes having the same indices as the point. So BP, which is perpendicular to OP, is parallel to the planes that are producing reflection P in Fig. 4.10a. If we draw a line parallel to *BP* and passing through *C,* the center of the circle, this line (or any other line parallel to it and separated from it by an integral multiple of d_{hkl}) represents a plane in the set that reflects the x-ray beam under these conditions. The beam impinges on this plane at the angle θ, is reflected at the same angle, and so diverges from the beam at *C* by the angle 2θ, which takes it precisely through point *P.* So *CP* gives the direction of the reflected ray *R* in Fig. 4.10*a.* In Fig. 4.10*b,* the reflected ray *R'* follows a different path, the line *CP'.*

The conclusion that reflection occurs in the direction *CP* when reciprocal-lattice point *P* comes in contact with this circle also holds for all points on all circles produced by rotating the circle of radius $1/\lambda$ about the x-ray beam. The figure that results, called the *sphere of reflection,* is shown in Fig. 4.11 intersecting the reciprocal-lattice planes *h0l* and *h1l.* In the crystal orientation shown, reciprocal-lattice point 012 is in contact with the sphere, so a diffracted ray *R* is diverging from the source beam in the direction defined by *C* and point 012. This ray would be detected as the 012 reflection.

As the crystal is rotated in the x ray beam, various reciprocal-lattice points come into contact with this sphere, each producing a beam in the direction of a line from the center of the sphere of reflection through the reciprocal-lattice point that is in contact with the sphere. The reflection produced when reciprocal-lattice point P_{hkl} contacts the sphere is called the *hkl* reflection and, according to Bragg's model, is caused by reflection from the set of equivalent, parallel, real-space planes (*hkl*).

This model of diffraction implies that the directions of reflection, as well as the number of reflections, depend only on unit-cell dimensions, and not on the contents of the unit cell. The intensity of reflection *hkl* depends on

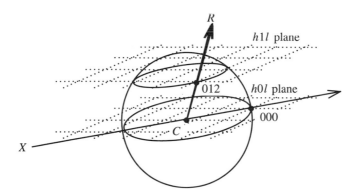

Figure 4.11 Sphere of reflection. When reciprocal-lattice point 012 intersects the sphere, ray *R* emerges from the crystal as reflection 012.

the values of $\rho(x,y,z)$ on planes (*hkl*). We will see (Chapter 5) that the intensities of the reflections give us the structural information we seek.

F. The number of measurable reflections

If the sphere of reflection has a radius of $1/\lambda$, then any reciprocal-lattice point within a distance $2/\lambda$ of the origin can be rotated into contact with the sphere of reflection (Fig. 4.12). This distance defines the *limiting sphere*. The number of reciprocal-lattice points within the limiting sphere is equal to the number of reflections that can be produced by rotating the crystal through all possible orientations in the x-ray beam. This demonstrates that the unit-cell dimensions and the wavelength of the x-rays determine the number of measurable reflections. Shorter wavelengths make a larger sphere of reflection, bringing more reflections into the measurable realm. Larger unit cells mean smaller reciprocal unit cells, which populate the limiting sphere more densely, also increasing the number of measurable reflections.

Because there is one lattice point per reciprocal unit cell (one-eighth of each lattice point lies within each of the eight unit-cell vertices), the number of reflections within the limiting sphere is approximately the number of reciprocal unit cells within this sphere. So the number N of possible reflections equals the volume of the limiting sphere divided by the volume V_{recip}

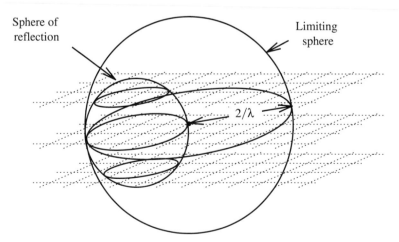

Figure 4.12 Limiting sphere. All reciprocal-lattice points within the limiting sphere of radius $2/\lambda$ can be rotated through the sphere of reflection.

of one reciprocal cell. The volume of a sphere of radius r is $(4\pi/3)r^3$, and r for the limiting sphere is $2/\lambda$, so

$$N = \frac{(4\pi/3) \cdot (2/\lambda)^3}{V_{recip}} \tag{4.6}$$

The volume V of the real unit cell is V_{recip}^{-1}, so

$$N = \frac{33.5 \cdot V}{\lambda^3} \tag{4.7}$$

Equation (4.7) shows that the number of available reflections depends only on V and λ. For a modest-size protein unit cell of dimensions $40 \times 60 \times 80$ Å, 1.54-Å radiation can produce 1.76×10^6 reflections, an overwhelming amount of data. Fortunately, because of cell and reciprocal-lattice symmetry, not all of these reflections are unique (Section III.G). Still, getting most of the available information from the diffraction experiment with protein crystals usually requires measuring somewhere between 10^3 and 10^6 reflections.

It can be further shown that the limit of resolution in an image derived from diffraction information is roughly equal to d_{min}, the minimum inter-planar spacing that gives a measurable reflection at the wavelength of the x-radiation. For instance, with 1.54-Å radiation, the resolution attainable from all the available data is 0.8 Å, which is more than needed to resolve atoms. A resolution of 1.5 Å, which barely resolves adjacent atoms, can be obtained from about half the available data. Interpretable electron-density maps can usually be obtained with data out to only 2.5 or 3 Å. The number of reflections out to 2.5 Å is roughly the volume of a limiting reciprocal sphere of radius $1/(2.5$ Å$)$ multiplied by the volume of the real unit cell. For the unit cell in the example above, this gives about 50,000 reflections. (For a sample calculation, see Chapter 8.)

G. Unit-cell dimensions

Because reciprocal-lattice spacings determine the angles of reflection, the spacings of reflections on the film are related to reciprocal-lattice spacings. (The exact relationship depends on the geometry of recording the reflec-tions, as discussed below.) Reciprocal-lattice spacings, in turn, are simply the inverse of real-lattice spacings. So the distances between reflections on the film and the dimensions of the unit cell are closely connected, making

it possible to measure unit-cell dimensions from film spacings. I will discuss the exact geometric relationship in Section III.F, in the context of data-
collection devices, whose geometry determines the method of computing
unit-cell size.

H. Unit-cell symmetry

If the unit-cell contents are symmetric, then the reciprocal lattice is also
symmetric and certain sets of reflections are equivalent. In theory, only one
member of each set of equivalent reflections need be measured, so awareness of unit-cell symmetry can greatly reduce the magnitude of data collection. In practice, modest redundancy of measurements improves accuracy,
so when more than one equivalent reflection is observed (measured), or
when the same reflection is observed more than once, the average of these
multiple observations is considered more accurate than any single observation.

In this section, I will discuss some of the simplest aspects of unit-cell
symmetry. Crystallography in practice requires detailed understanding of
these matters, but users of crystallographic models need only understand
their general importance. As we will see later (Section III.G, this chapter,
and Chapter 5), the crystallographer can determine the unit-cell symmetry
from a limited amount of x-ray data, and thus can devise a strategy for data
collection that will minimize repeated observation of equivalent reflections.

The symmetry of a unit cell is described by its space group, which is represented by a cryptic symbol (like $P\,2_12_12_1$), in which a capital letter indicates the lattice type and the other symbols represent symmetry operations
that can be carried out on the unit cell without changing its appearance.
Mathematicians in the late 1800s showed that there are exactly 230 possible
space groups.

The unit cells of a few lattice types are shown in Fig. 4.13. In this figure,
P designates a primitive lattice, containing one lattice point at each corner
or vertex of the cell. Because each lattice point is shared among eight
neighboring unit cells, a primitive lattice contains eight times one-eighth or
one lattice point per unit cell. Symbol I designates a body-centered or internal lattice, with an additional lattice point in the center of the cell, and
thus two lattice points per unit cell. Symbol F designates a face-centered
lattice, with additional lattice points (beyond the primitive ones) on the
center of each face.

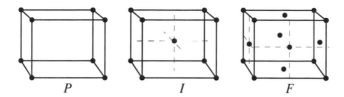

Figure 4.13 *P*, *I*, and *F* unit cells.

An example of a symmetry operation is rotation of an object about an axis. To illustrate with a familiar object, if a rectangular table is rotated 180° about an axis perpendicular to and centered on the tabletop (Fig. 4.14), the table looks just the same as it did before rotation (ignoring imperfections such as coffee stains). We say that the table possesses a twofold rotation axis, because in rotating the table one full circle about this axis, we find two positions that are equivalent: 0° and 180°. The axis itself is an example of a symmetry element.

Protein molecules are inherently asymmetric, being composed of chiral amino-acid residues coiled into larger chiral structures such as right-handed helices or twisted beta sheets. If only one protein molecule occupies a unit cell, then the cell itself is chiral, and there are no symmetry elements. This situation is rare; in most cases, the unit cell contains several identical molecules or oligomeric complexes in an arrangement that produces symmetry elements. In the unit cell, the largest aggregate of molecules that possesses no symmetry elements, but can be juxtaposed on other identical

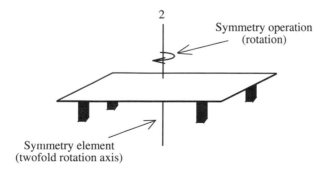

Figure 4.14 Table with a twofold axis of rotation.

entities by symmetry operations, is called the *asymmetric unit*. In the simplest case for proteins, the asymmetric unit is a single protein molecule.

The simplest symmetry operations and elements needed to describe unit-cell symmetry are translation, rotation (element: rotation axis), and reflection (element: mirror plane). Combinations of these elements produce more complex symmetry elements, including centers of symmetry, screw axes, and glide planes (below). Because proteins are inherently asymmetric, mirror planes and more complex elements involving them are not found in unit cells of proteins. All symmetry elements in protein crystals are translations, rotations, and screw axes, which are rotations and translations combined.

Translation simply means movement by a specified distance. For example, by the definition of *unit cell*, movement of its contents along one of the unit-cell axes by a distance equal to the length of that axis superimposes the atoms of the cell on corresponding atoms in the neighboring cell. This translation by one axial length is called a *unit translation*. Unit cells often exhibit symmetry elements that entail translations by a simple fraction of axial length, such as $a/4$.

In the space-group symbols, rotation axes such as the twofold axis of the table in Fig. 4.14 are represented in general by the symbol n, and specifically by a number. For example, 4 means a fourfold rotation axis. If the unit cell possesses this symmetry element, then it has the same appearance after each 90° rotation around the axis.

The screw axis results from a combination of rotation and translation. The symbol n_m represents an n-fold screw axis with a translation of m/n of the unit translation. For example, Plate 5 shows models of the amino acid alanine on a 3_1 screw axis in a hypothetical unit cell. On the screw axis, each successive molecule is rotated by 120° (360°/3) with respect to the previous one, and translated one-third of the axial length.

Plate 6 shows alanine in hypothetical unit cells of two space groups. A triclinic unit cell (Plate 6a) is designated $P1$, being a primitive lattice with only a onefold axis of symmetry (that is, with no symmetry). $P2_1$ (Plate 6b) describes a primitive unit cell possessing a twofold screw axis parallel to **c**, which points toward you as you view Plate 6. Notice that along any 2_1 screw axis, successive alanines are rotated 180° and translated one-half the axis length. A cell in space group $P2_12_12_1$ possesses three perpendicular twofold screw axes.

For the crystallographer, one of the most useful ways to describe unit-cell symmetry is by *equivalent positions*: positions in the unit cell that are superimposed on each other by the symmetry operations. In a $P2_1$ cell with an atom located at (x,y,z), an identical atom can be found at $(-x, -y, \frac{1}{2} + z)$, because the operation of a 2_1 screw axis interchanges these positions. So a $P2_1$ cell has the equivalent positions (x,y,z) and $(-x, -y, \frac{1}{2} + z)$. (The $\frac{1}{2}$

means one-half of a unit translation along $\mathbf{c}$, or a distance $c/2$ along the z-axis.)

Lists of equivalent positions for the 230 space groups can be found in *International Tables for X-ray Crystallography,* [1] a reference series that contains an enormous amount of practical information that crystallographers need in their daily work. So the easiest way to see how asymmetric units are arranged in a cell of complex symmetry is to look up the space group in *International Tables*. Each entry contains a list of equivalent positions for that space group and a diagram of the unit cell. The entry for space group $P2_1$ is shown in Fig. 4.15.

Certain symmetry elements in the unit cell announce themselves in the diffraction pattern by causing specific reflections to be missing (intensity of zero). For example, a twofold screw axis (2_1) along the $\mathbf{c}$ edge causes all $00l$ reflections having odd values of l to be missing. (Notice in Fig. 4.15 that "Conditions limiting possible reflections" in a $P2_1$ cell includes the condition that $l = 2n$, meaning that only the even-numbered reflections are present along the l-axis.) As another example, body-centered (I) lattices show missing reflections for all values of hkl where the sum of h, k, and l is odd. These patterns of missing reflections are called *systematic absences*, and they allow the crystallographer to determine the space group by looking at a few crucial planes of reflections. I will show later in this chapter how symmetry guides the strategy of data collection. In Chapter 5, I will show why symmetry causes systematic absences.

III. Collecting x-ray diffraction data

A. Introduction

Simply stated, the goal of data collection is to determine the indices and record the intensities of as many reflections as possible, as rapidly and efficiently as possible. One cause for urgency is that crystals, especially those of macromolecules, deteriorate in the beam because x-rays generate heat and reactive free radicals in the crystal. Thus the crystallographer would like to capture as many reflections as possible during every moment of irradiation. Often the diffracting power of the crystal limits the number of available reflections. Protein crystals that produce measurable reflections

---

[1] *International Tables for X-ray Crystallography*, Vol. 1, N. F. M. Henry and K. Lonsdale, eds., Reidel NE/ Kluwer Academic Publishers, Norwell, Massachusetts, 1969, p. 105.

$P2_1$
C_2^2

No. 4 $\qquad$ $P \, 1 \, 1 \, 2_1$ $\qquad$ 2 $\quad$ Monoclinic

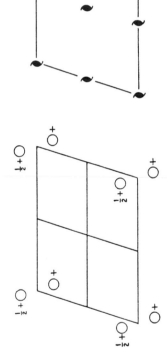

1ST SETTING $\qquad\qquad$ Origin on 2_1; unique axis c

Number of positions, Wyckoff notation and point symmetry			Co-ordinates of equivalent positions	Conditions limiting possible reflections
2	a	1	$x,y,z; \quad \bar{x},\bar{y},\tfrac{1}{2}+z.$	hkl: No conditions $hk0$: No conditions $00l$: $l=2n$

Symmetry of special projections

(001) $p2$; $\quad a'=a, \, b'=b$ $\qquad$ (100) $pg1$; $\quad b'=b, \, c'=c$ $\qquad$ (010) $p1g$; $\quad c'=c, \, a'=a$

Figure 4.15 $P2_1$ entry in *International Tables for X-ray Crystallography*, Vol. 1, N. F. M. Henry and K. Lonsdale, eds., Reidel NE/ Kluwer Academic Publishers, Norwell, Massachusetts, 1969, p. 105. Reprinted with permission of Reidel NE/Kluwer Academic Publishers.

from interplanar spacings down to about 3 Å or less are usually suitable for structure determination.

In the following sections, I will discuss briefly a few of the major instruments employed in data collection. These include the x-ray sources, which produce an intense, narrow beam of radiation; detectors, which allow quantitative measurement of reflection intensities; and cameras or diffractometers, which control the orientation of the crystal in the x-ray beam, and thus direct reflections having known indices to detectors.

B. X-ray sources

X-rays are electromagnetic radiation of wavelengths 0.1–100 Å. X-rays in the useful range for crystallography can be produced by bombarding a metal target (most commonly copper or molybdenum) with electrons produced by a heated filament and accelerated by an electric field. A high-energy electron collides with and displaces an electron from a low-lying orbital in a target metal atom. Then an electron from a higher orbital drops into the resulting vacancy, emitting its excess energy as an x-ray photon.

The element in the target exhibits narrow "characteristic lines" (specific wavelengths) of emission resulting from the characteristic energy-level spacing of that element. The wavelengths of emission lines are longer for elements of lower atomic number Z. For instance, electrons dropping from the L shell of copper ($Z = 29$) to replace displaced K electrons ($L \rightarrow K$ or K_α transition) emit x-rays of $\lambda = 1.54$ Å. The M $\rightarrow$ K transition produces a nearby emission band (K_β) at 1.39 Å (Fig. 4.16a, solid curve). For molybdenum ($Z = 42$), $\lambda(K_\alpha) = 0.71$ Å and $\lambda(K_\beta) = 0.63$ Å.

A monochromatic (single-wavelength) source of x-rays is desirable for crystallography because the diameter of the sphere of reflection is $1/\lambda$, and a source producing two distinct wavelengths of radiation gives two spheres of reflection and two interspersed sets of reflections, making indexing difficult or impossible because of overlapping reflections. Elements like copper and molybdenum make good x-ray sources if the weaker K_β radiation can be removed.

At wavelengths away from the characteristic emission lines, each element absorbs x-rays. The magnitude of absorption increases with increasing x-ray wavelength and then drops sharply just at the wavelength of K_β. The dashed curve in Fig. 4.16a shows the absorption spectrum for copper. The wavelength of this absorption edge, or sharp drop in absorption, like that of characteristic emission lines, increases as Z decreases such that the absorption edge for element $Z - 1$ lies slightly above the K_β emission line of element Z. This makes element $Z - 1$ an effective K_β filter for element Z,

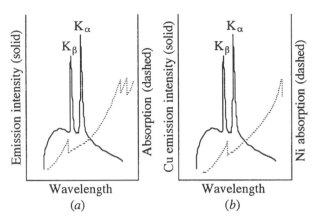

Figure 4.16 (*a*) Emission (solid) and absorption (dashed) spectra of copper. (*b*) Emission spectrum of copper (solid) and absorption spectrum of nickel (dashed). Notice that Ni absorbs K_β more strongly than K_α.

leaving almost pure monochromatic K_α radiation. For example, a nickel filter 0.015 mm in thickness reduces $Cu-K_\beta$ radiation to about 0.01 times the intensity of $Cu-K_\alpha$. Figure 4.16*b* shows the copper emission spectrum (solid line) and the nickel absorption spectrum (dashed line). Notice that Ni absorbs strongly at the wavelength of $Cu-K_\beta$ radiation, but transmits $Cu-K_\alpha$.

Copper is the most commonly used target for detection by film, which is more sensitive to $Cu-K_\alpha$ radiation than to that of molybdenum. Molybdenum, which gives a larger sphere of reflection, and hence the potential for better resolution, is more commonly used as a target when x-rays are detected by scintillation counting, as in diffractometry (see Section III.D of this chapter).

There are three common x-ray sources: *x-ray tubes* (actually a cathode-ray tube, sort of like a television tube), *rotating anode tubes,* and *particle accelerators*, which produce synchrotron radiation in the x-ray region. In the x-ray tube, electrons from a hot filament (cathode) are accelerated by electrically charged plates and collide with a water-cooled anode made of the target metal (Fig. 4.17*a*). X-rays are produced at low angles from the anode and emerge from the tube through windows of beryllium.

Output from x-ray tubes is limited by the amount of heat that can be dissipated from the anode by circulating water. Higher x-ray output can be obtained from rotating anode tubes, in which the target is a rapidly rotating metal

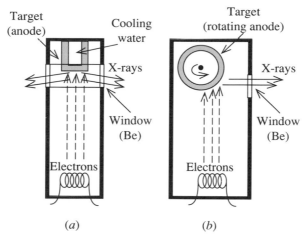

Figure 4.17 (*a*) X-ray tube. (*b*) Rotating anode tube.

disk (Fig. 4.17*b*). This arrangement improves heat dissipation by spreading the electron bombardment over a much larger piece of metal. Rotating anode sources are more than 10 times as powerful as tubes with fixed anodes.

Particle accelerators, which are used by physicists to study subatomic particles, are the most powerful x-ray sources. In these giant rings, electrons or positrons circulate at velocities near the speed of light, driven by energy from radio frequency transmitters, and maintained in circular motion by powerful magnets. A charged body like an electron emits energy (synchrotron radiation) when forced into curved motion, and in accelerators, the energy is emitted as x-rays. Accessory devices called "wigglers" cause additional bending of the beam, thus increasing the intensity of radiation. Systems of focusing mirrors and monochromators tangential to the storage ring provide powerful monochromatic x-rays at selectable wavelengths.

A photo and diagram of an accelerator, the Cornell Electron Storage Ring in Ithaca, New York, is shown in Fig. 4.18. The accelerator ring (white in photo and black ring in surrounding diagram) lies buried beneath a soccer field (cut away). The two large buildings lie on opposite sides of the field. Within the larger building is the Cornell High-Energy Synchrotron Source (CHESS), where synchrotron x-rays are provided at several work stations. Crystallographers can apply to CHESS for grants of time for data collection.

Although synchrotron sources are available only at particle accelerators, and require the crystallographer to collect data away from the usual site of

Figure 4.18 Cornell Electron Storage Ring. Photo and diagram reprinted with permission of Floyd R. Newman Laboratory of Nuclear Studies, Cornell University.

work, there are many advantages that compensate for the inconvenience. X-ray data that require several hours of exposure to a rotating anode source can often be obtained in seconds or minutes at a synchrotron source like CHESS. In two or three days at a synchrotron source, a crystallographer can collect data that might take months to acquire with conventional sources. Another advantage, as we will see in Chapter 6, is that x-rays of selectable wavelength can be helpful in solving the phase problem.

Whatever the source of x-rays, the beam is directed through a collimator, a narrow metal tube that selects and reflects the x-rays into parallel paths, producing a narrow beam. After collimation, beam diameter can be further reduced with systems of metal plates called *focusing mirrors*. In the ideal arrangement of source, collimators, and crystal, all points on the crystal can "see" through the collimator and mirrors to all line-of-sight points on the x-ray source.

X-ray sources pose some dangers that the crystallographer must consider in daily work. X-ray tubes require high-voltage power supplies containing large condensers that can produce a dangerous shock even after equipment is shut off. The x-rays themselves are relatively nonpenetrating but can cause serious damage to surface tissues. Even brief exposure to weak x-rays can damage eyes, so protective goggles are standard attire in the vicinity of x-ray sources. The direct beam is especially powerful, and its intensity is always reduced to a minimum during alignment of collimating mirrors or cameras. During data collection, the direct beam is blocked just beyond the crystal by a piece of metal called a "beam stop," which also has the beneficial effect of preventing excessive radiation from reaching the center of films or area detectors, thus obscuring low-angle reflections. In addition, the entire source, camera, and detector are usually surrounded by Plexiglas to block scattered radiation from the beam stop or collimators, but to allow observation of the equipment. As a check on the efficacy of measures to prevent x-ray exposure, the prudent crystallographer wears a dosage-measuring ring or badge during all work with x-ray equipment. These devices are periodically sent to radiation-safety labs for measurement of the x-ray dose received by the worker.

C. Detectors

As mentioned previously, the simplest x-ray detector is x-ray-sensitive film. Various types of cameras (next section) can direct reflections to films in useful arrangements, allowing precise determination of indices and intensities for thousands of reflections on a single film. Scanners can be programmed to index these films and to measure the optical absorbance of each reflection, producing a computer file of indexed intensities.

In the use of film, there is an interesting technical problem with a simple solution. A single x-ray photon carries enough energy to expose a single grain of silver in the film, so spot density is proportional to the number of photons until many grains in a specific area are exposed. After this point, there is high probability that a photon will strike an already-exposed grain. As a result, film response becomes nonlinear for strong reflections. In order to measure the full range of reflection intensities, the crystallographer uses two or more films in a stack. The weaker reflections will be recorded satisfactorily on the first film. Stronger reflections saturate the first film, but they reach the second film attenuated by passage through the first. Each reflection will thus be recorded accurately on at least one film in the stack. By correcting for the (known) attenuation of overlying films, the crystallographer can obtain accurate x-ray intensities over a wider range than a single film would allow.

Reflection intensities can also be measured by scintillation counters, which in essence count the x-ray photons, and thus give quite accurate intensities over a wide range. Scintillation counters contain a material that produces a flash of light (a scintillation) when it absorbs an x-ray photon. A photocell counts the flashes. With simple scintillation counters, each reflection must be measured separately, an arrangement that is convenient only in diffractometry (next section).

Modern area detectors combine the accuracy and wide range of scintillation counting with the simultaneous measurement of many reflections, as on film. As an example, the Mark I detector at the University of California, San Diego, as diagrammed in Fig. 4.19, consists of two perpendicular sets of parallel wires in a flat box filled with an inert gas. A window of beryllium permits entry of x-rays from the front of the detector.

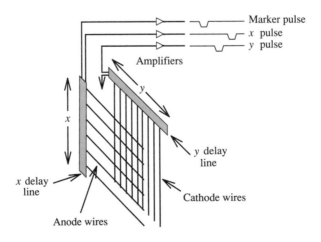

Figure 4.19 Exploded view of detector wires in multiwire area detector.

One set of wires, say the x set in Fig. 4.19, is the anode, while the perpendicular set is the cathode. The anode set is held at a high positive voltage relative to the cathode set. The wires of each set terminate in a delay line, which delays any signal from a wire in its set by a time interval proportional to the distance of the wire from the end of the delay line. This time delay allows determination of which wire produced the signal.

Entering the detector through the beryllium window, an x-ray photon ionizes the gas in a small region ($\sim$100 μm), producing a few hundred electrons. The electrons drift to the nearest anode wire and, because of the high voltage, each electron triggers an electrical discharge that in turn produces thousands of ion pairs in the gas. The movement of these ions in the electric field of the cathode and anode wires produces a pulse of current in each of the nearest wires. The detection of these pulses at ends of the x and y delay lines allows determination of the reflection position in the detector. A small pulse that appears instantly in the ground connection of the anode delay line serves as a marker against which to time the x and y pulses.

Pulses appear in several parallel lines in both the x and y sets, with the strongest pulses in the lines nearest the initial ionization. The known characteristic shape of such a pulse can be fitted to the pulse heights from several neighboring lines, thus locating the ionization event to a higher resolution than that of the wire spacing in the detector.

The output from the area detector is fed to a computer, which indexes the event using the x and y positional information and the crystal orientation at the time of the event. The computer sums events that have the same index, and thus produces a file of indexed intensities.

D. Diffractometers and cameras

Between the irradiated crystal and the detector lies a device for precisely orienting the crystal so as to direct specific reflections toward the detector. In this section, I will describe some of these devices, including the diffractometer, which directs single reflections to a scintillation counter or a larger number of reflections to an area detector; and several cameras, which direct large numbers of reflections simultaneously to film or area detectors. In all cases, the task of these devices is to rotate a crystal through a series of known orientations, causing specified reciprocal-lattice points to pass through the sphere of reflection and thus produce diffracted x-ray beams.

In all forms of data collection, the crystal is mounted on a goniometer head, which allows the crystallographer to set the crystal orientation precisely. The goniometer head (Fig. 4.20) consists of a holder for a capillary tube containing the crystal; two arcs (marked by angle scales), which permit rotation of the crystal by 40° in each of two perpendicular planes; and two dovetailed sledges, which permit small translations of the arcs for centering the crystal on the rotation axis of the head.

Protein crystals are usually sealed in capillary tubes with the mother liquor, and then mounted on the goniometer head, which is adjusted to center one face of the crystal perpendicular to the x-ray beam, and to allow rotation of the crystal while maintaining centering.

The initial crystal orientations or "settings" are determined by physical examination. Well-formed crystals show distinct faces that are parallel to unit-cell edges, and first attempts to obtain a diffraction pattern are made by placing a crystal face perpendicular to the x-ray beam. Preliminary study of diffraction patterns allows determination of unit-cell dimensions and internal symmetry, as described below. Using this information, the crystallographer decides on the best set of crystal orientations for collecting data

Figure 4.20 Goniometer head, with capillary tube holder at top. The tool (right) is an Allen wrench for adjusting arcs and sledges. Photo courtesy of Charles Supper Company.

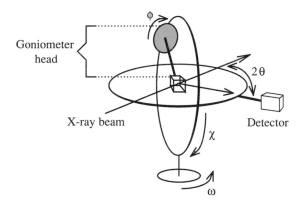

Figure 4.21 System of circles in diffractometry. The crystal in the center is mounted on a goniometer head.

and uses the external faces of the crystal as a guide for orienting unit-cell axes.

In the diffractometer, the goniometer head and crystal are mounted in a system of movable circles called a *goniostat*, which allows automated movement of the crystal into almost any orientation with respect to the x-ray beam and the detector. See Figs. 4.21 and 4.22.

The complete diffractometer consists of a fixed x-ray source, the goniostat, and a movable scintillation-counter detector. The system of circles (Fig. 4.21) allows rotation of the goniometer head (angle ϕ), movement of the head around a circle centered on the x-ray beam (angle χ), and rotation of the χ circle around an axis perpendicular to the beam (angle ω). Furthermore, the detector moves on a circle coplanar with the beam. The axis of this circle coincides with the ω-axis. The position of the detector with respect to the beam is denoted by the angle 2θ. With this arrangement, the crystal can be moved to bring any reciprocal-lattice point that lies within the limiting sphere into the plane of the detector and into contact with the sphere of reflection, producing diffracted rays in the detector plane. The detector can be moved into proper position to receive, and measure the intensity of, the resulting diffracted beam.

Modern diffractometers are computer driven and almost completely automated. They can, with some minimal (but important) intervention by the operator, search for reflections and determine unit-cell dimensions, and then systematically measure the intensities of all accessible reflections.

This kind of diffractometry gives highly accurate intensity measurement but is slow in comparison with methods that record many reflections at once. In addition, the total irradiation time is long, so crystals may deteriorate and have to be replaced. While one reflection is being recorded, there are usually other unmeasured reflections present, so a considerable amount of diffracted radiation is wasted. More recently, diffractometers have been teamed up with area detectors, as shown in Fig. 4.22, giving substantial increases in the efficiency of data collection.

In this photo, the goniostat (*a*) and area detector (*c*) are separated by a drum of helium (*b*) that transmits x-rays with less loss than air. The crystal (*d*) is barely visible in the tiny glass tube. The two arrows (*e*) mark the collimator (left) and the beam stop (right), which prevents the direct x-ray beam from reaching the detector. Arrows 1, 2, and 3 on the goniostat indicate the χ, ϕ, and ω circles.

Of the many types of x-ray cameras, only two are still in wide use, the Buerger or precession camera, and the Weissenberg or rotation-oscillation

Figure 4.22 Diffractometer and area detector. Photo courtesy of Professor Leonard J. Banaszak.

camera. The precession camera is used (frequently with film) primarily in preliminary studies to determine crystal quality, unit cell dimensions, and symmetry, as well as to assess the quality of derivative crystals. The rotation-oscillation camera is used with film or area detectors to measure large numbers of reflections simultaneously.

The precession camera (Fig. 4.23), although the more complicated in its motion, produces the simplest diffraction pattern. X-rays enter through the black tube at the left to strike the crystal, mounted in a goniometer head. Beyond the crystal are an annular-screen holder (smaller black square) and a film holder (larger black square). The remaining machinery moves crystal, screen, and film in a precessing motion about the x-ray beam.

Precession photographs reveal the reflections in an undistorted image of the reciprocal lattice. Figure 4.24a shows the geometry that allows precession photography. If the crystal is aligned with one of the real-space axes (a so-called direct axis, as opposed to a reciprocal axis) parallel to the

Figure 4.23 Precession camera with mounted goniometer head. To see a precession photograph, refer to Fig. 2.6. Photo courtesy of Charles Supper Company.

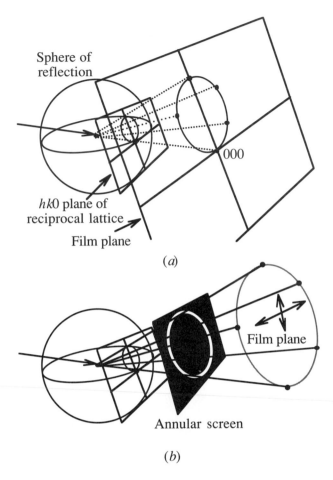

Figure 4.24 (a) Geometry of precession photography. Reflections from a single plane of the reciprocal lattice emerge from the crystal on the surface of a cone. (b) Annular screen selects one cone of reflections, and thus one plane of reciprocal-lattice points, for photography.

beam, then the zero-level reciprocal-lattice plane along that axis is tangent to the sphere of reflection at the origin. For example, in an orthorhombic system with the **c** axis (and hence also the **c*** axis) parallel to the beam, the $hk0$ plane is tangent to the sphere at the origin. If the crystal is then inclined slightly, as in Fig. 4.24, this zero-level lattice intersects the sphere of reflection in a circle (small circle in the figure), and any reciprocal-lattice points on this circle produce reflections whose paths lie on a cone radiating from the center of the sphere of reflection through the intersecting circle. If the crystal then precesses about the beam axis at this same angle of inclination, and the film is made to precess about the beam axis in the same manner, the zero-level reflections ($hk0$ in this case) will fall on the film in an undistorted projection of their spatial relationship in the reciprocal lattice. Of course, other non-zero-level points will also intersect the sphere, but their reflections emerge on smaller or larger cones and can be eliminated by interposing an annular screen between crystal and film (Fig. 4.24b). This screen also precesses so that its open annulus always transmits the cone of zero-level reflections. The result is a diffraction pattern recorded as in Chapter 2, Fig. 2.6.

Indexing the resulting photograph is straightforward. In our orthorhombic system, with c^* precessing about the beam, if a^* lies along the x-axis and b^* along y, then the reflection directly to the right of the origin is the 100 reflection, and the one just above the origin is 010.

The crystal is first aligned visually. If the crystal is not perfectly aligned with direct axis precessing about the beam, the image of the reciprocal lattice will be distorted. Precise measurements of the distortion provide information that allows the alignment to be corrected. The resulting undistorted image of the reciprocal lattice allows simple measurement of unit-cell dimensions, as discussed in the next section.

Precession cameras are complex, but give the diffraction pattern in its simplest, most understandable form. Rotation-oscillation cameras are far simpler, merely providing means to rotate the crystal about an axis perpendicular to the beam, as well as to oscillate it back and forth by a few degrees about the same axis. This movement casts large numbers of reflections in a complex pattern onto the film or area detector (Fig. 4.25).

To produce this figure, a computer program calculated the indices of reflections expected during an oscillation photograph in which the crystal is oscillated about its c axis. At the expected position of each reflection, the program can plot the indices of that reflection. Only the l index of each reflection is shown here, revealing that reflections from many levels of reciprocal space are recorded at once. Although oscillation photographs are very complex, once the unit-cell dimensions are known, all reflections can be indexed.

As the crystal oscillates about a fixed starting position, a limited number of reciprocal-lattice points pass back and forth through the sphere of reflection, and their intensities are recorded. The amount of data from a single oscillation range is limited only by overlap of reflections. The strategy is to take photographs by oscillating the crystal through a small angle about a

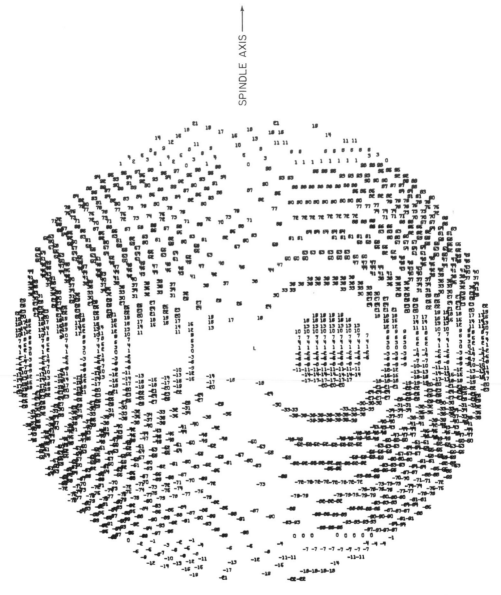

Figure 4.25 Diagram showing expected positions of reflections in an oscillation photograph. Diagram courtesy of Professor Michael Rossmann.

starting position of rotation, recording all the resulting reflections, then rotate the crystal to a new starting point such that the new oscillating range
overlaps the previous one slightly. From this new position, oscillation produces additional reflections. This process is continued until all reflections
have been recorded.

Oscillation photography can be used in tandem with area detectors, but
detector resolution limits the size of the oscillation angle. Large unit
cells, such as those of virus capsids, mean small reciprocal unit cells and
closely spaced reflections. Film is commonly used in such cases, because
its greater spatial resolution allows more reflections to be captured in each
oscillation.

E. Scaling and postrefinement of intensity data

The goal of data collection is a set of consistently measured, indexed intensities for as many of the reflections as possible. After data collection, the
raw intensities must be processed to improve their consistency and to maximize the number of measurements that are sufficiently accurate to be used.

A complete set of measured intensities often includes distinct blocks of
data obtained from several (or many) crystals, and if data are collected on
film, from many films. Because of variability in diffracting power of crystals, intensity of the x-ray beam, and sensitivity of films (if used), the crystallographer cannot assume that the absolute intensities are consistent from
one block of data to the next. An obvious way to obtain this consistency is
to compare reflections of the same index that were measured from more
than one crystal or on more than one film, and rescale the intensities of the
two blocks of data so that identical reflections are given identical intensities. This process is called *scaling*. With films, scaling is often preliminary
to a more complex process, *postrefinement*, which recovers usable data
from reflections that were only partially measured.

Primarily because real crystals are mosaics of submicroscopic crystals
(Chapter 3, Section I.B), a reciprocal-lattice point acts as a small three-dimensional entity (sphere or ovoid) rather than as an infinitesimal point. As
a reciprocal-lattice point moves through the sphere, diffraction is weak at
first, peaks when the center of the point lies precisely on the sphere, and
then weakens again before it is extinguished. Accurate measurement of intensity thus entails recording the x-ray output during the entire passage of
the point through the sphere. Any range of oscillation will record some reflections only partially, but these may be recorded fully at another rotation

angle, allowing partial reflections to be discarded from the data. The problem of partial reflections is serious for large unit cells, where smaller oscillation angles are employed to minimize overlap of reflections. In such cases, if partial reflections are discarded, a great deal of data is lost.

Data from partial reflections can be interpreted accurately through post-refinement of the intensity data. This process produces an estimate of the partiality of each reflection. Partiality is a fraction p $(0 > p > 1)$ that can be used as a correction factor to convert the measured intensity of a partial reflection to an estimate of that reflection's full intensity.

Scaling and postrefinement are the final stages in producing a list of internally consistent intensities for most of the available reflections.

F. Determining unit cell dimensions

The unit-cell dimensions determine the reciprocal-lattice dimensions, which in turn tell us where we must look for the data. Methods like oscillation photography require that we know precisely which reflections will fall completely and partially within a given oscillation angle, so we can collect as many reflections as possible without overlap. So the unit-cell dimensions are needed to devise a strategy of data collection that will give us as many identifiable (by index), measurable reflections as possible.

Diffractometer software can search for reflections, measure their precise positions, and subsequently compute unit-cell parameters. This search entails complexities we need not encounter here. Instead, I will illustrate the simplest method for determining unit-cell dimensions: measuring reflection spacings from an orthorhombic crystal on a precession photograph.

As discussed above, a precession photograph is an undistorted projection of the reciprocal-lattice points onto a flat film. Because reciprocal-lattice spacings are the inverse of real-lattice spacings, the unit-cell dimensions are inversely proportional to the spacing of reflections on a precession photograph. Figure 4.26 shows the geometric relationship between reflection spacings on the film and actual reciprocal-lattice spacings.

The crystal is precessing about its $\mathbf{c}^*$ axis, which records $hk0$ reflections on the film, with the $h00$ axis horizontal and the $0k0$ axis vertical. Point P is the reciprocal-lattice point 100, in contact with the sphere of reflection, and O is the origin. Point F is the origin on the film and R is the recording of reflection 100 on the film. The distance OP is the reciprocal of the distance d_{100}, which is the length of unit cell edge $\mathbf{a}$. Because CRF and CPO

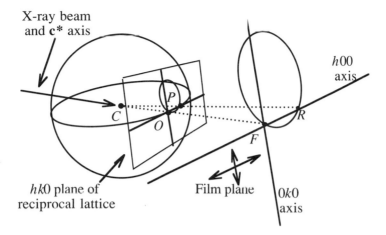

Figure 4.26 Reflection spacings on the film are directly proportional to reciprocal-lattice spacings, and so they are inversely proportional to unit cell dimensions.

are similar triangles (all corresponding angles equal), and because the radius of the sphere of reflection is $1/\lambda$,

$$\frac{RF}{CF} = \frac{PO}{CO} = \frac{PO}{(1/\lambda)} = PO \cdot \lambda \tag{4.8}$$

Therefore,

$$PO = \frac{RF}{CF \cdot \lambda} \tag{4.9}$$

Because $d_{100} = 1/PO$,

$$d_{100} = \frac{CF \cdot \lambda}{RF} \tag{4.10}$$

In other words, the axial length a (length of unit cell edge **a**) can be determined by dividing the crystal-to-film distance (CF) by the distance from the film origin to the 100 reflection (RF) and multiplying the quotient by the wavelength of x-rays used in taking the photograph.

In like manner, the vertical reflection spacing along $0k0$ or parallel axes gives $1/d_{010}$, and from it, the length of unit-cell axis **b**. A second precession photograph, taken after rotating this orthorhombic crystal by 90° about its vertical axis, would record the 00l axis horizontally, giving $1/d_{001}$ and the length of **c**.

Of course, the distance from the film origin to the 100 reflection on a precession photograph is the same as the distance between any two reflections along this or other horizontal lines, so one photograph allows many measurements to determine accurately the *average* spacing of reciprocal-lattice points along two different axes. From accurate average values, unit-cell-axis lengths can be determined with sufficient accuracy to guide a data-collection strategy.

G. Symmetry and the strategy of collecting data

Strategy of data collection is guided not only by the unit cell's dimensions but also by its internal symmetry. If the cell and its contents are highly symmetric, then certain sets of crystal orientations produce exactly the same reflections, reducing the number of crystal orientations needed in order to obtain all the distinct or unique reflections.

As mentioned earlier, the unit-cell space group can be determined from systematic absences in the diffraction pattern. With the space group in hand, the crystallographer can determine the space group of the reciprocal lattice, and thus know which orientations of the crystal will give identical data. All reciprocal lattices possess a symmetry element called a *center of symmetry* or *point of inversion* at the origin. That is, the intensity of each reflection *hkl* is identical to the intensity of reflection *–h –k –l*. To see why, recall from our discussion of lattice indices (Section II.B) that the the index of the (230) planes can also be expressed as (–2 –3 0). In fact, the 230 and the –2 –3 0 reflections come from opposite sides of the same set of planes, and the reflection intensities are identical. (The equivalence of I_{hkl} and I_{-h-k-l} is called *Friedel's law*, but there are exceptions. See Anomalous scattering, Chapter 6, Section IV.) This means that half of the reflections in the reciprocal lattice are redundant, and data collection that covers 180° about any reciprocal-lattice axis will capture all unique reflections.

Additional symmetry elements in the reciprocal lattice allow further reduction in the total angle of data collection. It can be shown that the reciprocal lattice possesses the same symmetry elements as the unit cell, plus the additional point of inversion at the origin. The 230 possible space groups reduce to only 11 different groups, called *Laue groups*, when a center of symmetry is added. For each Laue group, and thus for all reciprocal lattices, it is possible to compute the fraction of reflections that are unique. For monoclinic systems, the center of symmetry is the only element added in the reciprocal lattice and the fraction of unique reflections is $\frac{1}{4}$. At the other extreme, for the cubic space group P_{432}, which possesses four-, three-, and twofold rotation axes, only $\frac{1}{48}$ of the reflections are unique. Determination of the crystal symmetry can greatly reduce the number of reflections that must be measured. It also guides the crystallographer in choosing the best axis about which to rotate the crystal during data collection.

IV. Summary

The result of x-ray data collection is a list of intensities, each assigned indices *hkl* corresponding to its position in the reciprocal lattice. The intensity assigned to reflection *hkl* is therefore a measure of the relative strength of the reflection from the set of lattice planes having indices *hkl*. Recall that indices are counted from the origin (indices 000), which lies in the direct path of the x-ray beam. In an undistorted image of the reciprocal lattice, such as a precession photograph (or its equivalent computed from diffractometer or oscillation data), reflections having low indices lie near the origin, and those with high indices lie farther away. Also recall that as indices increase, there is a corresponding decrease in the spacing d_{hkl} of the real-space planes represented by the indices. This means that the reflections near the origin come from sets of widely spaced planes, and thus carry information about larger features of the molecules in the unit cell. On the other hand, the reflections far from the origin come from closely spaced lattice planes in the crystal, and thus they carry information about the fine details of structure. In the next three chapters, I will examine the relationship between the intensities of the reflections and the molecular structures we seek, and thus show how the crystallographer extracts structural information from the list of intensities.

5 From Diffraction Data to Electron Density

I. Introduction

In producing an image of molecules from crystallographic data, the computer simulates the action of a lens, computing the electron density within the unit cell from the list of indexed intensities obtained by the methods described in Chapter 4. In this chapter, I will discuss the mathematical relationships between the crystallographic data and the electron density.

As I stated in Chapter 2, computation of the Fourier transform is the lens-simulating operation that a computer performs to produce an image of molecules in the crystal. The Fourier transform describes precisely the mathematical relationship between an object and its diffraction pattern. The transform allows us to convert a Fourier-series description of the reflections to a Fourier-series description of the electron density. A reflection can be described by a Fourier series, called a *structure-factor equation*, containing one term for each atom (or each volume element) in the unit cell. In turn, the electron density is described by a Fourier series in which each term is a structure factor. The crystallographer uses the Fourier transform to convert the structure factors to $\rho(x,y,z)$, the desired electron-density equation.

First I will discuss Fourier series and the Fourier transform in general terms. I will emphasize the form of these equations and the information they contain, in hopes of helping you interpret the equations, that is, translate the equations into words and visual images. Then I will present the specific types of Fourier series that represent structure factors and electron density, and show how the Fourier transform interconverts them.

II. Fourier series and the Fourier transform

A. One-dimensional waves

Recall from Chapter 2, Section VI.A, that waves are described by periodic functions, and that simple wave equations can be written in the form

$$f(x) = F\cos 2\pi(hx + \alpha) \tag{5.1}$$

or

$$f(x) = F\sin 2\pi(hx + \alpha) \tag{5.2}$$

where $f(x)$ specifies the vertical height of the wave at any horizontal position x (measured in wavelengths, where $x = 1$ implies one full wavelength or one full repeat of the periodic function). In these equations, F specifies the amplitude of the wave (its height from peak to valley), h specifies its frequency (number of wavelengths per radian), and α specifies its phase (position of the wave, in radians, with respect to the origin). These equations are *one*-dimensional in the sense that they represent a numerical value [$f(x)$, the height of the wave] at all points along *one* axis, in this case, the x-axis. See Fig. 2.13 for graphs of such equations.

I also stated in Chapter 2 that any wave, no matter how complicated, can be described as the sum of simple waves. This sum is called a Fourier "series," and each simple wave equation in the series is called a Fourier "term." Either of Equations (5.1) or (5.2) could be used as single Fourier terms. For example, we can write a Fourier series of n terms using Equation (5.1) as follows:

$$f(x) = F_0 \cos 2\pi (0x + \alpha_0)$$
$$+$$
$$F_1 \cos 2\pi (1x + \alpha_1)$$
$$+$$
$$F_2 \cos 2\pi (2x + \alpha_2) \qquad (5.3)$$
$$+$$
$$\cdots$$
$$+$$
$$F_n \cos 2\pi (nx + \alpha_n)$$

or equivalently

$$f(x) = \sum_{h=0}^{n} F_h \cos 2\pi (hx + \alpha_h) \qquad (5.4)$$

According to Fourier theory, any complicated periodic function can be approximated by this series. Think of the cosine terms as basic waveforms that can be used to build any other waveform. Also according to Fourier theory, we can use the sine function, or for that matter, *any* periodic function, in the same way, as the basic waveform for building any other periodic function.

A very useful basic waveform is $\lfloor\cos 2\pi(hx) + i \sin 2\pi(hx)\rfloor$. Here, the waveforms of cosine and sine are combined to make a complex number, whose general form is $a + ib$, where i is the imaginary number $(-1)^{1/2}$. Although the phase α of this waveform is not shown, it is implicit in the combination of the cosine and sine functions, and it depends only on the values of h and x. As I will show in Chapter 6, expressing a Fourier term in this manner gives a clear geometric means of representing the phase α and allows us to see how phases are computed. For now, just accept this convention as a convenient way to write completely general Fourier terms. In Chapter 6, I will discuss the properties of complex numbers and show how they are used to represent and compute phases.

With the terms written in this fashion, a Fourier series looks like this:

$$f(x) = \sum_{h=0}^{n} F_h [\cos 2\pi (hx) + i \sin 2\pi (hx)] \qquad (5.5)$$

In words, this series is the sum of n simple Fourier terms, one for each integral value of h beginning with zero and ending with n. Each term is a simple wave with its own amplitude F_h, its own frequency h, and (implicitly) its own phase α_h.

Next, we can express the complex number in square brackets as an exponential, using this equality from complex number theory:

$$\cos \theta + i \sin \theta = e^{i\theta} \tag{5.6}$$

In our case, $\theta = 2\pi(hx)$, so the Fourier series becomes

$$f(x) = \sum_{h=0}^{n} F_h e^{2\pi i(hx)} \tag{5.7}$$

or simply

$$f(x) = \sum_{h} F_h e^{2\pi i(hx)} \tag{5.8}$$

in which the sum is taken over all values of h, and the number of terms is unspecified.

I will write Fourier series in this form throughout the remainder of the book. This kind of equation is compact and handy but quite opaque at first encounter. Take the time now to look at this equation carefully and think about what it represents. Whenever you see an equation like this, just remember that it is a Fourier series, a sum of sine and cosine wave equations, with the full sum representing some complicated wave. The hth term in the series, $F_h e^{2\pi i(hx)}$, can be expanded to $F_h [\cos 2\pi(hx) + i \sin 2\pi(hx)]$, making plain that the hth term is a simple wave of amplitude F_h, frequency h, and implicit phase α_h.

B. Three-dimensional waves

The Fourier series that the crystallographer seeks is $\rho(x,y,z)$, the three-dimensional electron density of the molecules under study. This function is a wave equation or periodic function because it repeats itself in every unit cell. The waves described in the equations above are one-dimensional; they represent a numerical value $f(x)$ that varies in one direction, along the x-axis. How do we write the equations of two-dimensional and three-dimensional waves? First, what do the graphs of such waves look like?

When you graph a function, you must use one more dimension than specified by the function. You use the additional dimension to represent the numerical value of the function. For example, in graphing $f(x)$, you use the

y-axis to show the numerical value of $f(x)$. [In Fig. 2.13, the y-axes are used to represent $f(x)$, the height of each wave at point x.] Graphing a two-dimensional function $f(x,y)$ requires the third dimension to represent the numerical value of the function.

For example, imagine a weather map with mountains whose height at location (x,y) represents the temperature at that location. Such a map graphs a two-dimensional function $t(x,y)$, which gives the temperature t at all locations (x,y) on the plane represented by the map. If we must avoid using the third dimension, for instance in order to print a flat map, the best we can do is to draw a contour map on the plane map (Fig. 5.1), with continuous lines (contours, in this case called *isotherms*) representing locations having the same temperature.

Graphing the three-dimensional function $\rho(x,y,z)$ in the same manner would require four dimensions, one for each of the spatial dimensions x, y, and z, and a fourth one for representing the value of ρ. Here a contour map is the only choice. In three dimensions, contours are continuous surfaces (rather than lines) on which the function has a constant numerical value. A contour map of the three-dimensional wave $\rho(x,y,z)$ exhibits surfaces of

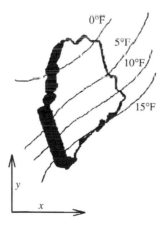

Figure 5.1 Seasonable February morning in Maine. Lines of constant temperature (isotherms) allow plotting a two-dimensional function without using the third dimension. This is a contour map of $t(x,y)$, giving the temperature t at all locations (x,y). Along each contour line lie all points having the same temperature. A planar contour map of a function of two variables takes the form of contour lines on the plane. In contrast, a contour map of a function of three variables takes the form of contour surfaces in three dimensions (see Plate 2).

constant electron density ρ. You are already familiar with such contour maps. The common drawings of electronic orbitals (such as the $1s$ orbital of a hydrogen atom, often drawn as a simple sphere) is a contour map of a three-dimensional function. Everywhere on the surface of this sphere, the electron density is the same. Orbital surfaces are often drawn to enclose the region that contains 98% (or some specified value) of the total electron density.

The blue netlike surface in Plate 2 is also a contour map of a three-dimensional function. It represents a surface on which the electron density $\rho(x,y,z)$ of adipocyte lipid-binding protein (ALBP) is constant. Imagine that the net encloses 98% (or some specified value) of the protein's electron density, and so the net is in essence an image of the protein's surface.

I hope the foregoing helps you to imagine three-dimensional waves. What do the equations of such waves look like? A three-dimensional wave has three frequencies, one along each of the x-, y-, and z-axes. So three variables h, k, and l are needed to specify the frequency in each of the three directions. A general Fourier series for the wave $f(x,y,z)$, written in the compact form of Equation (5.8) is as follows:

$$f(x,y,z) = \sum_h \sum_k \sum_l F_{hkl}\, e^{2\pi i(hx + ky + lz)} \tag{5.9}$$

In words, Equation (5.9) says that the complicated three-dimensional wave $f(x,y,z)$ can be represented by a Fourier series. Each term in the series is a simple three-dimensional wave whose frequency is h in the x direction, k in the y direction, and l in the z direction. For each possible set of values h, k, and l, the associated wave has amplitude F_{hkl}, and implicitly, phase α_{hkl}. The triple sum simply means to add up terms for all possible sets of integers h, k, and l. The range of values for h, k, and l depends on how many terms are required to represent the complicated wave $f(x,y,z)$ to the desired precision.

C. The Fourier transform: General features

Fourier demonstrated that for any function $f(x)$, there exists another function $F(h)$ such that

$$F(h) = \int_{-\infty}^{+\infty} f(x)\, e^{2\pi i(hx)}\, dx \tag{5.10}$$

where $F(h)$ is called the *Fourier transform* (FT) of $f(x)$, and the units of the variable h are reciprocals of the units of x. For example, if x is time in seconds (s), then h is reciprocal time, or frequency, in reciprocal seconds (s^{-1}). So if $f(x)$ is a function of time, $F(h)$ is a function of frequency. Taking the FT of time-dependent functions is a means of decomposing these functions into their component frequencies and is sometimes referred to as *Fourier analysis*. The FT in this form is used in infrared (IR) and nuclear magnetic resonance (NMR) spectroscopy to obtain the frequencies of many spectral lines simultaneously.

On the other hand, if x is a length in Å, h is reciprocal length in $Å^{-1}$. You can thus see that this highly general mathematical form is naturally adapted for relating real and reciprocal space. In fact, as I mentioned earlier, the Fourier transform is a precise mathematical description of diffraction. The diffraction patterns in Figs. 2.7–2.10 are Fourier transforms of the corresponding simple objects and arrays. If these figures give you some intuition about how an object is related to its diffraction pattern, they provide the same perception about the kinship between an object and its Fourier transform.

According to Equation (5.10), to compute $F(h)$, the Fourier transform of $f(x)$, just multiply the function by $e^{2\pi i (hx)}$ and integrate (or better, let a computer integrate) the combined functions with respect to x. The result is a new function $F(h)$, which is the FT of $f(x)$. Computer programs for calculating FTs of functions are widely available.

The Fourier transform operation is reversible. That is, the same mathematical operation that gives $F(h)$ from $f(x)$ can be carried out in the opposite direction, to give $f(x)$ from $F(h)$; specifically

$$f(x) = \int_{-\infty}^{+\infty} F(h)\, e^{-2\pi i (hx)}\, dh \qquad (5.11)$$

In other words, if $F(h)$ is the transform of $f(x)$, then $f(x)$ is in turn the transform of $F(h)$. [In this situation, $f(x)$ is sometimes called the *back-transform* of $F(h)$, but this is a loose term that simply refers to the second successive transform that re-creates the original function.] Notice that the only difference between Equations (5.10) and (5.11) is the sign of the exponential term. You can think of this sign change as analogous (very *roughly* analogous!) to the sign change that makes subtraction the reverse of addition. Adding 3 to 5 gives 8: $5 + 3 = 8$. To reverse the operation and generate the original 5, you subtract 3 from the previous result: $8 - 3 = 5$. If you think of 8 as a simple transform of 5 made by adding 3, the back-transform of 8 is 5, produced by subtracting 3.

Returning to the visual transforms of Figs. 2.7–2.10, each object on the left (the sphere, for instance) is the Fourier transform (the back-transform, if you wish) of its diffraction pattern. If we build a model that looks like the diffraction pattern on the right, and then obtain its diffraction pattern, we get an image of the object on the left.

One added complication: The functions $f(x)$ and $F(h)$ above are one-dimensional. Fortunately, the Fourier transform applies to periodic functions in any number of dimensions. To restate Fourier's conclusion in three dimensions, for any function $f(x,y,z)$ there exists the function $F(h,k,l)$ such that

$$F(h,k,l) \;=\; \int_x \int_y \int_z f(x,y,z)\, e^{2\pi i(hx+ky+lz)}\, dx\, dy\, dz \qquad (5.12)$$

As before, $F(h,k,l)$ is called the Fourier transform of $f(x,y,z)$, and in turn, $f(x,y,z)$ is the Fourier transform of $F(h,k,l)$ as follows:

$$f(x,y,z) \;=\; \int_h \int_k \int_l F(h,k,l)\, e^{-2\pi i(hx+ky+lz)}\, dh\, dk\, dl \qquad (5.13)$$

Thinking again about the potential usefulness of computing FTs in crystallography, you will see that we can use the Fourier transform to obtain information about real space, $f(x,y,z)$, from information about reciprocal space, $F(h,k,l)$. Specifically, the diffraction pattern contains information whose Fourier transform is information about the contents of the unit cell.

D. Fourier this and Fourier that: Review

I have used Fourier's name in discussing several types of equations and operations, and I want to be sure that I have not muddled them in your mind. First, a *Fourier series* is a sum of simple wave equations or periodic functions that describes or approximates a complicated periodic function. Second, constructing a Fourier series, that is, determining the proper F, h, and α values to approximate a specific function, is called *Fourier synthesis*. For example, the sum of f_0 through f_6 in Fig. 2.14 is at once a Fourier series and the product of Fourier synthesis. Third, decomposing a complicated function into its components is called *Fourier analysis*. Fourth and finally, the

Fourier transform is an operation that transforms a function containing variables of one type (say, time) into a function whose variables are reciprocals of the original type [in this case, $1/(\text{time})$ or frequency]. The function $f(x)$ is related to its Fourier transform $F(h)$ by Equation (5.10). The term *transform* is commonly used as a noun to refer to the function $F(h)$, and also loosely as a verb to denote the operation of computing a Fourier transform. (Last, a simple bit of grammatical awkwardness: The word *series* is both singular and plural. You must gather from context whether a writer is talking about one series or many series.)

III. Fourier mathematics and diffraction

A. Structure factor as a Fourier series

I have stated that both structure factors and electron density can be expressed as Fourier series. A structure factor describes one diffracted x-ray, which produces one reflection received at the detector. A structure factor F_{hkl} can be written as a Fourier series in which each term gives the contribution of one atom to the reflection hkl [see Fig. 2.15 and Equation (2.3)]. Here is a single term, called an *atomic structure factor* f_{hkl}, in such a series, representing the contribution of the single atom j to reflection hkl:

$$f_{hkl} = f_j\, e^{2\pi i(hx_j + ky_j + lz_j)} \tag{5.14}$$

The term f_j is called the *scattering factor* of atom j, and it is a mathematical function (called a δ function) that amounts to treating the atom as a simple sphere of electron density. The function is slightly different for each element, because each element has a different number of electrons (a different value of Z) to diffract the x-rays. The exponential term should be familiar to you by now. It represents a simple three-dimensional periodic function having both cosine and sine components. But the terms in parentheses now possess added physical meaning: x_j, y_j, and z_j are the coordinates of atom j in the unit cell (real space), expressed as fractions of the unit-cell axis lengths; and h, k, and l, in addition to their role as frequencies of a wave in the three directions x, y, and z, are also the indices of a specific reflection in the reciprocal lattice.

As mentioned earlier, the phase of a wave is implicit in the exponential formulation of a structure factor and depends only on the atomic coordinates (x_j, y_j, z_j) of the atom. In fact, the phase for diffraction by one atom is $2\pi(hx_j + ky_j + lz_j)$, the exponent of e (ignoring the imaginary i) in the structure factor. For its contribution to the 220 reflection, an atom at $(0, \frac{1}{2}, 0)$ has phase $2\pi(hx_j + ky_j + lz_j)$ or $2\pi(2[0] + 2[\frac{1}{2}] + 0[0]) = 2\pi$, which is the same as a phase of zero. This atom lies on the (220) plane, and all atoms lying on (220) planes contribute to the 220 reflection with phase of zero. [Try the preceding calculation for another atom at $(\frac{1}{2}, 0, 0)$, which is also on a (220) plane.] This is in keeping with Bragg's law, which says that all atoms on a set of equivalent, parallel lattice planes diffract in phase with each other.

Each diffracted ray is a complicated wave, the sum of diffractive contributions from *all* atoms in the unit cell. For a unit cell containing n atoms, the structure factor F_{hkl} is the sum of all the atomic f_{hkl} values for individual atoms. Thus, in parallel with Equation (2.3), we write the structure factor for reflection F_{hkl} as follows:

$$F_{hkl} = \sum_{j=1}^{n} f_j e^{2\pi i (hx_j + ky_j + lz_j)} \tag{5.15}$$

In words, the structure factor that describes reflection hkl is a Fourier series in which each term is the contribution of one atom, treated as a simple sphere of electron density. So the contribution of each atom j to F_{hkl} depends on (1) what element it is, which determines f_j, the amplitude of the contribution and (2) its position in the unit cell (x_j, y_j, z_j), which establishes the phase of its contribution.

Alternatively, F_{hkl} can be written as the sum of contributions from each volume element of electron density in the unit cell [see Fig. 2.16 and Equation (2.4)]. The electron density of a volume element centered at (x, y, z) is, roughly, the average value of $\rho(x, y, z)$ in that region. The smaller we make our volume elements, the more precisely these averages approach the correct values of $\rho(x, y, z)$ at all points. We can, in effect, make our volume elements infinitesimally small, with the average values of $\rho(x, y, z)$ precisely equal to the actual values at every point, by integrating the function $\rho(x, y, z)$ rather than summing average values. Think of the resulting integral as the sum of the contributions of an infinite number of vanishingly small volume elements. Written this way,

$$F_{hkl} = \iiint_{h\ k\ l} \rho(x, y, z) \, e^{2\pi i (hx + ky + lz)} \, dx \, dy \, dz \tag{5.16}$$

or equivalently

$$F_{hkl} = \int_V \rho\,(x,y,z)\,e^{2\pi i\,(hx+ky+lz)}\,dV \tag{5.17}$$

where the integral over V, the unit-cell volume, is just shorthand for the integral over all values of x, y, and z in the unit cell. Each volume element contributes to F_{hkl} with a phase determined by its coordinates (x,y,z), just as the phase of atomic contributions depend on atomic coordinates.

We can see by comparing Equation (5.17) with Equation (5.10) [or (5.16) with (5.12)] that F_{hkl} is the Fourier transform of $\rho(x,y,z)$. More precisely, F_{hkl} is the transform of $\rho(x,y,z)$ on the set of real-lattice planes (hkl). All of the F_{hkl}'s together comprise the transform of $\rho(x,y,z)$ on all sets of equivalent, parallel planes throughout the unit cell.

B. Electron density as a Fourier series

Because the Fourier transform operation is reversible [Equations (5.10) and (5.11)], the electron density is in turn the transform of the structure factors, as follows:

$$\rho\,(x,y,z) = \frac{1}{V}\sum_h\sum_k\sum_l F_{hkl}\,e^{-2\pi i\,(hx+ky+lz)} \tag{5.18}$$

where V is the volume of the unit cell.

This transform is a triple sum rather than a triple integral because the F_{hkl}'s represent a set of discrete entities: the reflections of the diffraction pattern. The transform of a discrete function, such as the reciprocal lattice of measured intensities, is a summation of discrete values of the function. The transform of a continuous function, such as $\rho(x,y,z)$, is an integral, which you can think of as a sum also, but a sum of an infinite number of infinitesimals.

Superficially, except for the sign change (in the exponential term) that accompanies the transform operation, this equation appears identical to Equation (5.9), a general three-dimensional Fourier series. But here, each F_{hkl} is not just one of many simple numerical amplitudes for a standard set of component waves in a Fourier series. Instead, each F_{hkl} is a structure factor, itself a Fourier series, describing a specific reflection in the diffraction pattern. ("Curiouser and curiouser," said Alice.)

C. Computing electron density from data

Equation (5.18) tells us, at last, how to obtain $\rho(x,y,z)$. We need merely to construct a Fourier series from the structure factors. The structure factors describe diffracted rays that produce the measured reflections. A full description of a diffracted ray, like any description of a wave, must include three parameters: amplitude, frequency, and phase. In discussing data collection, however, I mentioned only two measurements: the indices of each reflection and its intensity. Looking again at Equation (5.18), you see that the indices of a reflection play the role of the three frequencies in one Fourier term. The only measurable variable remaining in the equation is F_{hkl}. Does the measured intensity of a reflection, the only measurement we can make in addition to the indices, completely define F_{hkl}? Unfortunately, the answer is "no."

D. The phase problem

Because F_{hkl} is a periodic function, it possesses amplitude, frequency, and phase. It represents a diffracted x-ray, so its frequency is that of the x-ray source. The amplitude of F_{hkl} is proportional to the square root of I_{hkl}, so structure amplitudes are directly obtainable from measured reflection intensities. But the phase of F_{hkl} is not directly obtainable from a single measurement of the reflection intensity. In order to compute $\rho(x,y,z)$ from the structure factors, we must obtain, in addition to the intensity of each reflection, the phase of each diffracted ray. In Chapter 6, I will present an expression for $\rho(x,y,z)$ as a Fourier series in which the phases are explicit, and I will discuss means of obtaining phases. For now, on the assumption that the phases can be obtained, and thus that complete structure factors are obtainable, I will consider further the implications of Equations (5.15) (F's in terms of atoms), (5.16) [F's in terms of $\rho(x,y,z)$], and (5.18) [$\rho(x,y,z)$ in terms of F's].

IV. The meaning of the Fourier equations

A. Reflections as Fourier terms: Equation (5.18)

First consider Equation (5.18) (ρ in terms of F's). Each term in this Fourier-series description of $\rho(x,y,z)$ is a structure factor representing a single x-ray reflection. The indices hkl of the reflection give the three frequencies necessary to describe the Fourier term as a simple wave in three dimensions. Recall from Chapter 2, Section VI.B, that any periodic function can be approximated by a Fourier series, and that the approximation improves as more terms are added to the series (see Fig. 2.14). The low-frequency terms in Equation (5.18) determine gross features of the periodic function $\rho(x,y,z)$, while the high-frequency terms improve the approximation by filling in fine details. You can also see in Equation (5.18) that the low-frequency terms in the Fourier series that describes our desired function $\rho(x,y,z)$ are given by reflections with low indices, that is, by reflections near the center of the diffraction pattern (Fig. 5.2).

The high-frequency terms are given by reflections with high indices, reflections farthest from the center of the pattern. Thus you can see the importance of how well a crystal diffracts. If a crystal does not produce diffracted rays at large angles from the direct beam (reflections with large indices), the Fourier series constructed from all the measurable reflections lacks high-frequency terms, and the resulting transform is not highly detailed—the resolution of the resulting image is poor. The Fourier series of Fig. 2.14 is truncated in just this manner and does not fit the target function in fine details like the sharp corners.

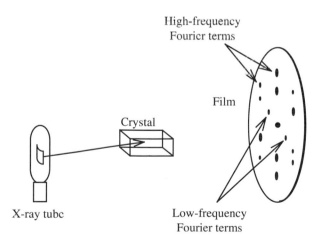

Figure 5.2 Structure factors of reflections near the center of the diffraction pattern are low-frequency terms in the Fourier series that approximates $\rho(x,y,z)$. Structure factors of reflections near the edge of the pattern are high-frequency terms.

B. Computing structure factors from a model: Equations (5.15) and (5.16)

Equation (5.15) describes one structure factor in terms of diffractive contributions from all atoms in the unit cell. Equation (5.16) describes one structure factor in terms of diffractive contributions from all volume elements of electron density in the unit cell. These equations suggest that we can calculate all of the structure factors either from an atomic model of the protein or from an electron-density function. In short, if we know the structure, we can calculate the diffraction pattern, including the phases of all reflections. This computation, of course, appears to go in just the opposite direction that the crystallographer desires. It turns out, however, that computing structure factors from a model of the unit-cell contents is an essential part of crystallography, for several reasons.

First, this computation is used in obtaining phases. As I will discuss in Chapter 6, the crystallographer obtains phases by starting from rough estimates of them, and then undertaking an iterative process to improve the estimates. This iteration entails two alternating steps: (1) computing an estimated $\rho(x,y,z)$ (i.e., a crude model of the structure), using Equation (5.18) with observed intensities (I_{obs}) and estimated phases (α_{calc}), and then (2) computing new structure factors (F_{calc}), using either Equation (5.16) with the crude model of $\rho(x,y,z)$ from step 1, or Equation (5.15) with a partial atomic model of the molecule, containing only those atoms that can be definitely located. The second computation produces a new set of estimated phases, and the cycle is repeated: a new estimated $\rho(x,y,z)$ (a more complete model) is used to compute new phases. With each cycle, the crystallographer hopes to obtain better phases, and better estimates of $\rho(x,y,z)$, which means more detailed electron-density maps and more complete and accurate models of the desired structure.

I will discuss the iterative improvement of phases and electron-density maps in Chapter 7. For now, just take note that obtaining the final structure entails calculating both $\rho(x,y,z)$ from structure factors, and structure factors from $\rho(x,y,z)$.

Equations (5.15) and (5.16) also provide a means of monitoring the iterative process to see whether it is converging toward improved phases and improved $\rho(x,y,z)$. The computed structure factors F_{calc} include both the desired phases α_{calc} and a new set of intensities. I will refer to these *calculated* intensities as I_{calc} to distinguish them from the *measured* reflection intensities I_{obs}. As the iteration proceeds, the values of I_{calc} should approach those of I_{obs}. So the crystallographer compares the I_{calc} and I_{obs} values at each cycle in order to see whether the iteration is converging. When cycles of computation provide no further improvement in correspondence between calculated and measured intensities, then the process is complete.

C. Systematic absences in the diffraction pattern: Equation (5.15)

Finally, Equation (5.15) allows us to understand how systematic absences in the diffraction pattern reveal symmetry elements in the unit cell, thus guiding the crystallographer in assigning the space group of the crystal. Recall from Chapter 4, Section II.H, that if the unit cell possesses symmetry elements, then certain sets of reciprocal-lattice points are equivalent, and so certain reflections in the diffraction pattern are redundant. The crystallographer must determine the unit-cell space group (i.e., determine what symmetry elements are present) in order to devise an efficient strategy for measuring as many unique reflections as possible. I stated without justification in Chapter 4 that certain symmetry elements announce themselves in the diffraction pattern as "systematic absences": regular patterns of missing reflections. Now I will use Equation (5.15) to show how a symmetry element in the unit cell produces systematic absences in the diffraction pattern.

As an example, if the **c** axis of the unit cell is a twofold screw axis, then reflections 001, 003, 005, and all other 00l reflections in which l is an odd number are missing. We can see why by using the concept of equivalent positions (Chapter 4, Section II.H). For a unit cell with a twofold screw axis along edge **c**, the equivalent positions are (x, y, z) and $(-x, -y, z + \frac{1}{2})$. That is, for every atom j with coordinates (x, y, z) in the unit cell, there is an identical atom j' at $(-x, -y, z + \frac{1}{2})$. Atoms j and j' are called *symmetry-related atoms*. According to Equation (5.15), the structure factor for reflections F_{00l} is

$$F_{00l} = \sum_j f_j e^{2\pi i (l z_j)} \tag{5.19}$$

The exponential term is greatly simplified in comparison to that in Equation (5.15) because $h = k = 0$ for reflections on the 00l axis. Now I will separate the contributions of atoms j from their symmetry-related atoms j' :

$$F_{00l} = \sum_j f_j e^{2\pi i (l z_j)} + \sum_{j'} f_{j'} e^{2\pi i (l z_{j'})} \tag{5.20}$$

Because atoms j and j' are identical, I can substitute f_j for $f_{j'}$ and factor out the f terms:

$$F_{00l} = \sum_j f_j \left(\sum_j e^{2\pi i l z_j} + \sum_{j'} e^{2\pi i l z_{j'}} \right) \tag{5.21}$$

If the z coordinate of atom j is z, then the z coordinate of atom j' is $z + \frac{1}{2}$. Making these substitutions for z_j and $z_{j'}$,

$$F_{00l} = \sum_j f_j \left(\sum_j \left[e^{2\pi i l z} + e^{2\pi i l (z + 1/2)} \right] \right) \tag{5.22}$$

The f_j terms are nonzero, so F_{00l} is zero and the corresponding $00l$ reflection is missing only if all terms summed in brackets equal zero. Simplifying one of these terms,

$$e^{2\pi i l z} + e^{2\pi i l (z + 1/2)} = e^{2\pi i l z} (1 + e^{\pi i l}) \tag{5.23}$$

This term is zero, and hence F_{00l} is zero, if $e^{\pi i l}$ is -1. Converting this exponential to its trigonometric form,

$$e^{\pi i l} = \cos(\pi l) + i \sin(\pi l) \tag{5.24}$$

The cosine of π radians (180°), or any odd multiple of π radians, is -1. The sine of π radians is 0. Thus $e^{\pi i l}$ equals -1 for all odd values of l, and F_{00l} equals zero if l is odd.

The preceding shows that F_{00l} disappears for odd values of l when the **c** edge of a unit cell is a twofold screw axis. But what is going on physically? In short, the diffracted rays from two atoms at (x,y,z) and $(-x, -y, z + \frac{1}{2})$ are identical in amplitude ($f_j = f_{j'}$) but precisely opposite in phase. Thus the pair of atoms contributes nothing to F_{00l} when l is odd. Putting it another way, if the unit cell contains a twofold screw axis along edge **c**, then every atom in the unit cell is paired with a symmetry-related atom that cancels its contributions to all odd-numbered $00l$ reflections.

Similar computations have been carried out for all symmetry elements and combinations of elements. Like equivalent positions, systematic absences are tabulated for all space groups in *International Tables*, so the crystallographer can use this reference as an aid to space-group determination. The *International Tables* entry for space group $P2_1$ (Fig. 4.15), which possesses a 2_1 axis on edge **c**, shows that for reflections $00l$ the "Conditions limiting possible reflections" are $l = 2n$. In other words, in this space group, reflections $00l$ are present only if l is even (2 times any integer n), so they are absent if l is odd, as proved above.

V. Summary: From data to density

When we describe structure factors and electron density as Fourier series, we find that they are intimately related. The electron density is the Fourier transform of the structure factors, which means that we can convert the crystallographic data into an image of the unit cell and its contents. One necessary piece of information is, however, missing for each structure factor. We can measure only the intensity I_{hkl} of each reflection, not the complete structure factor F_{hkl}. What is the relationship between them? It can be shown that the amplitude of structure factor F_{hkl} is proportional to $(I_{hkl})^{1/2}$, the square root of the measured intensity. So if we know I_{hkl} from diffraction data, we know the amplitude of F_{hkl}. Unfortuantely, we do not know its phase α_{hkl}. In focusing light reflected from an object, a lens maintains all phase relationships among the rays, and thus constructs an image accurately. When we record diffraction intensities, we lose the phase information that the computer needs in order to simulate an x-ray-focusing lens. In Chapter 6, I will consider how to learn the phase of each reflection and thus to obtain the complete structure factors needed to calculate the electron density.

6

Obtaining Phases

I. Introduction

The molecular image that the crystallographer seeks is a contour map of the electron density $\rho(x,y,z)$ throughout the unit cell. The electron density, like all periodic functions, can be represented by a Fourier series. The representation that connects $\rho(x,y,z)$ to the diffraction pattern is

$$\rho(x,y,z) = \frac{1}{V} \sum_h \sum_k \sum_l F_{hkl} e^{-2\pi i(hx + ky + lz)} \qquad (5.18)$$

Equation (5.18) tells us how to calculate $\rho(x,y,z)$: simply construct a Fourier series using the structure factors F_{hkl}. For each term in the series, h, k, and l are the indices of reflection hkl, and F_{hkl} is the structure factor that describes the reflection. Each structure factor F_{hkl} is a complete description of a diffracted ray recorded as reflection hkl. Because it is a wave equation, F_{hkl} must specify frequency, amplitude, and phase. Its frequency is that of the x-ray source. Its amplitude is proportional to $(I_{hkl})^{1/2}$, the square root of the measured intensity I_{hkl} of reflection hkl. Its phase is unknown and is the only additional information the crystallographer needs in order to compute $\rho(x,y,z)$ and thus obtain an image of the protein. In this chapter, I will discuss some of the common methods of obtaining phases. Let me emphasize that each reflection has a phase, and so this phase problem must be solved for each one of the thousands of reflections used to construct the Fourier series that approximates $\rho(x,y,z)$.

In order to illuminate both the phase problem and its solution, I will represent structure factors as vectors on a two-dimensional plane of complex numbers of the form $a + ib$, where i is the imaginary number $(-1)^{1/2}$. This allows me to show geometrically how to compute phases. I will begin by introducing complex numbers and their representation as points having coordinates (a,b) on the complex plane. Then I will show how to represent structure factors as vectors on the same plane. Because we will now start thinking of the structure factor as a vector, I will hereafter write it in boldface ($\mathbf{F}_{hkl}$) instead of the italics used for simple variables and functions. Finally, I will use the vector representation of structure factors to explain a few common methods of obtaining phases.

II. Two-dimensional representation of structure factors

A. Complex numbers in two dimensions

Complex numbers of the form $N = a + ib$, where $i = (-1)^{1/2}$, can be represented as points in two dimensions (Fig. 6.1).

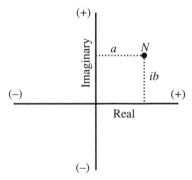

Figure 6.1 The complex number $N = a + ib$, represented as a point on the plane of complex numbers.

The horizontal axis in the figure represents the real-number line. Any real number a is a point on this line, which stretches from $-\infty$ to $+\infty$. The vertical axis is the imaginary-number line, on which lie all imaginary numbers ib between $-i\infty$ and $+i\infty$. A complex number $a + ib$, which possesses both real (a) and imaginary (ib) parts, is thus a point at position (a,b) on this plane.

B. Structure factors as complex vectors

A representation of structure factors on this plane must include the two properties we need in order to construct $\rho(x,y,z)$: amplitude and phase. Crystallographers represent each structure factor as a *complex vector*, that is, a vector (not a point) on the plane of complex numbers. The length of this vector represents the amplitude of the structure factor. Thus the length of the vector representing structure factor $\mathbf{F}_{hkl}$ is proportional to $(I_{hkl})^{1/2}$. The second property, phase, is represented by the angle α that the vector makes with the real-number line when the origin of the vector is placed at the origin of the complex plane, the point $0 + i0$. See Fig. 6.2a

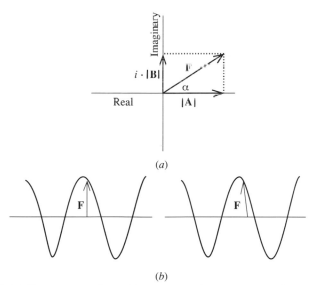

(a)

(b)

Figure 6.2 (a) The structure factor $\mathbf{F}$, represented as a vector on the plane of complex numbers. The length of $\mathbf{F}$ is proportional to $I^{1/2}$, the square root of the measured intensity I, and the angle between $\mathbf{F}$ and the real axis is the phase α. (b) (Stereo pair) $\mathbf{F}$ can be pictured as a complex vector spinning around its line of travel. The projection of the path taken by the head of the vector is the familiar sine wave.

We can represent a structure factor $\mathbf{F}$ as a vector $\mathbf{A} + i\mathbf{B}$ on this plane. The projection of $\mathbf{F}$ on the real axis is its real part $\mathbf{A}$, a vector of length $|\mathbf{A}|$ on the real-number line; and the projection of $\mathbf{F}$ on the imaginary axis is its imaginary part $i\mathbf{B}$, a vector of length $|\mathbf{B}|$ on the imaginary-number line. The length or magnitude (or in wave terminology, the amplitude) of a complex vector is analogous to the absolute value of a real number, so the length of vector $\mathbf{F}_{hkl}$ is $|\mathbf{F}_{hkl}|$; therefore, $|\mathbf{F}_{hkl}|$ is proportional to $(I_{hkl})^{1/2}$, and if the intensity is known from data collection, we can treat $|\mathbf{F}_{hkl}|$ as a known quantity. The angle that $\mathbf{F}_{hkl}$ makes with the real axis is represented in radians as α ($0 \leq \alpha \leq 2\pi$), or in cycles as α' ($0 \leq \alpha' \leq 1$), and is referred to as the *phase angle*.

This representation of a structure factor is equivalent to thinking of a wave as a complex vector spinning around its axis as it travels through space (Fig. 6.2b). If its line of travel is perpendicular to the tail of the vector, then a projection of the head of the vector along the line of travel is the familiar sine wave. The phase of a structure factor tells us the position of the vector at some arbitrary origin, and to know the phase of all reflections means to know all their phase angles with respect to a common origin.

In Chapter 4, Section III.G, I mentioned Friedel's law, that $I_{hkl} = I_{-h-k-l}$. It will be helpful for later discussions to look at the vector representations of pairs of structure factors $\mathbf{F}_{hkl}$ and $\mathbf{F}_{-h-k-l}$, which are called *Friedel pairs*. While I_{hkl} and I_{-h-k-l} are equal, $\mathbf{F}_{hkl}$ and $\mathbf{F}_{-h-k-l}$ are not. The structure factors of Friedel pairs have opposite phases, as shown in Fig. 6.3. This means that $\mathbf{F}_{-h-k-l}$ is the mirror image of $\mathbf{F}_{hkl}$ with the real axis serving as the mirror. Another way to put it is that Friedel pairs are reflections of each other in the real axis.

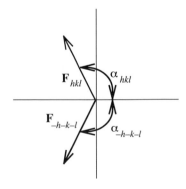

Figure 6.3 Structure factors of a Friedel pair: $\mathbf{F}_{-h-k-l}$ is the reflection of $\mathbf{F}_{hkl}$ in the real axis.

The representation of structure factors as vectors in the complex plane (or complex vectors) is useful in several ways. Because the diffractive contributions of atoms or volume elements to a single reflection are additive, each contribution can be represented as a complex vector, and the resulting structure factor is the vector sum of all contributions. For example, in Fig. 6.4, **F** represents a structure factor of a three-atom structure, in which f_1, f_2, and f_3 are the atomic structure factors.

The length of each atomic structure factor **f** represents its amplitude, and its angle α_n with the real axis represents its phase. The vector sum $\mathbf{F} = \mathbf{f}_1 + \mathbf{f}_2 + \mathbf{f}_3$ is obtained by placing the tail of $\mathbf{f}_1$ at the origin, the tail of $\mathbf{f}_2$ on the head of $\mathbf{f}_1$, and the tail of $\mathbf{f}_3$ on the head of $\mathbf{f}_2$, all the while maintaining the phase angle of each vector. The structure factor **F** is thus a vector with its tail at the origin and its head on the head of $\mathbf{f}_3$. This process sums both amplitudes and phases, so the resultant length of **F** represents its amplitude, and the resultant angle α is its phase angle. (The atomic vectors may be added in any order with the same result.)

In subsequent sections of this chapter, I will use this simple vector arithmetic to show how to compute phases from various kinds of data. In the next section, I will use complex vectors to derive an equation for electron density as a function of reflection intensities and phases.

C. Electron density as a function of intensities and phases

Figure 6.2 shows how to decompose $\mathbf{F}_{hkl}$ into its amplitude $|\mathbf{F}_{hkl}|$, which is the length of the vector, and its phase α_{hkl}, which is the angle the vector

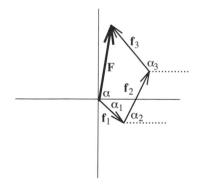

Figure 6.4 Molecular structure factor **F** is the vector sum of three atomic structure factors. Vector addition of $\mathbf{f}_1$, $\mathbf{f}_2$, and $\mathbf{f}_3$ gives the amplitude and phase of **F**.

makes with the real-number line. This allows us to express $\rho(x,y,z)$ as a function of the measurable amplitude of $\mathbf{F}$ (measurable because it can be computed from the reflection intensity I) and the unknown phase α. For clarity, I will at times drop the subscripts on $\mathbf{F}$, I, and α, but remember that these relationships hold for all reflections. In Fig. 6.2,

$$\cos\alpha = \frac{|\mathbf{A}|}{|\mathbf{F}|} \quad \text{and} \quad \sin\alpha = \frac{|\mathbf{B}|}{|\mathbf{F}|} \tag{6.1}$$

and therefore

$$|\mathbf{A}| = |\mathbf{F}| \cdot \cos\alpha \quad \text{and} \quad |\mathbf{B}| = |\mathbf{F}| \cdot \sin\alpha \tag{6.2}$$

Expressing $\mathbf{F}$ as a complex vector $\mathbf{A} + i\mathbf{B}$,

$$\mathbf{F} = |\mathbf{A}| + i|\mathbf{B}| = |\mathbf{F}| \cdot (\cos\alpha + i\sin\alpha) \tag{6.3}$$

Expressing the complex term in the parentheses as an exponential [Equation (5.6)],

$$\mathbf{F} = |\mathbf{F}| \cdot e^{i\alpha} \tag{6.4}$$

Substituting this expression for F_{hkl} in Equation (5.18), the electron-density equation (remembering that α is the phase α_{hkl} of a specific reflection), gives

$$\rho(x,y,z) = \frac{1}{V} \sum_h \sum_k \sum_l |\mathbf{F}_{hkl}| e^{i\alpha_{hkl}} e^{-2\pi i(hx + ky + lz)} \tag{6.5}$$

We can combine the exponential terms more simply by expressing the phase angle as α', using $\alpha = 2\pi\alpha'$:

$$\rho(x,y,z) = \frac{1}{V} \sum_h \sum_k \sum_l |\mathbf{F}_{hkl}| e^{2\pi i\alpha'_{hkl}} e^{-2\pi i(hx + ky + lz)} \tag{6.6}$$

Now we can combine the exponentials by adding their exponents:

$$\rho(x,y,z) = \frac{1}{V} \sum_h \sum_k \sum_l |\mathbf{F}_{hkl}| e^{-2\pi i(hx + ky + lz - \alpha'_{hkl})} \tag{6.7}$$

This equation gives the desired electron density as a function of the known amplitudes $|\mathbf{F}|$ and the unknown phases α'_{hkl} of each reflection. Recall that this equation represents $\rho(x,y,z)$ in a now-familiar form, as a Fourier series, but this time with the phase of each structure factor expressed explicitly. Each term in the series is a three-dimensional wave of amplitude $|\mathbf{F}_{hkl}|$, phase α'_{hkl}, and frequencies h along the x-axis, k along the y-axis, and l along the z-axis.

The most demanding element of macromolecular crystallography is the so-called phase problem, which involves determining the phase angle α_{hkl} for each reflection. In the remainder of this chapter, I will discuss some of the common methods for overcoming this obstacle. These include the *heavy-atom method* (also called *isomorphous replacement*), *anomalous scattering* (also called *anomalous dispersion*), and *molecular replacement*. Each of these techniques yields only estimates of phases, which must be improved before an interpretable electron-density map can be obtained. In addition, these techniques usually yield estimates for a limited number of the phases, so phase determination must be extended to include as many reflections as possible. In Chapter 7, I will discuss methods of phase improvement and phase extension, which ultimately result in accurate phases and an interpretable electron-density map.

III. The heavy-atom method (isomorphous replacement)

Each atom in the unit cell contributes to every reflection in the diffraction pattern [Equation (5.15)]. The contribution of an atom is greatest to the reflections whose indices correspond to lattice planes that intersect that atom, so a specific atom contributes to some reflections strongly, and to some weakly or not at all. If we could add one or a very small number of atoms to identical sites in all unit cells of a crystal, we would expect to see changes in the diffraction pattern, as the result of the additional contributions of the added atom. As I will show below, the slight perturbation in the diffraction pattern caused by an added atom can be used to obtain initial estimates of phases. In order for these perturbations to be large enough to measure, the added atom must be a strong diffractor, which means it must be an element of high atomic number, a so-called heavy atom.

A. Preparing heavy-atom derivatives

After obtaining a complete set of x-ray data, and determining that these data are adequate to produce a high-resolution structure, the crystallographer undertakes to prepare one or more heavy-atom derivatives. In the most common technique, crystals of the protein are soaked in solutions of heavy ions, for instance ions or ionic complexes of Hg, Pt, or Au. In many cases, such ions bind to one or a few specific sites on the protein without perturbing its conformation or crystal packing. For instance, surface cysteine residues react readily with Hg^{2+} ions, and cysteine, histidine, and methionine displace chloride from Pt complexes like $PtCl_4^{2-}$ to form stable Pt adducts. The conditions that give such specific binding must be found by simply trying different ionic compounds at various pH values and concentrations.

Several diffraction criteria define a promising heavy-atom derivative. First, the derivative crystals must be *isomorphic* with native crystals. At the molecular level, this means that the heavy atom must not disturb crystal packing or conformation of the protein. Unit-cell dimensions are quite sensitive to such disturbances, so heavy-atom derivatives whose unit-cell dimensions are the same as those of native crystals are probably isomorphous. The term *isomorphous replacement* comes from this criterion.

The second criterion for useful heavy-atom derivatives is that there must be measurable changes in at least a modest number of reflection intensities. These changes are the handle by which phase estimates are pulled from the data, so they must be clearly detectable, and large enough to measure accurately.

Figure 6.5 shows precession photographs for native and derivative crystals of the MoFe protein of nitrogenase. Underlined in the figure are pairs of reflections whose relative intensities are altered by the heavy atom. In examining heavy-atom photos by eye, the crystallographer looks for pairs of reflections whose relative intensities are reversed. This distinguishes real heavy-atom perturbations from simple differences in overall intensity of two photos. For example, consider the leftmost underlined pairs in each photograph. In the native photo (*a*), the reflection on the right is the darker of the pair, while in the derivative photo (*b*), the reflection on the left is darker. Several additional differences suggest that this derivative might produce good phases.

Finally, the derivative crystal must diffract to reasonably high resolution, although the resolution of derivative data need not be as high as that of native data. Methods of phase extension (Chapter 7) can produce phases for higher-angle reflections from good phases of reflections at lower angles.

Having obtained a suitable derivative, the crystallographer faces data collection again. Since derivatives must be isomorphous with native crystals, the strategy is the same as for collecting the native data. You can see that the phase problem effectively multiplies the magnitude of the crystallographic project by the number of derivative data sets needed. As I will show, at least two, and often more, derivatives are required.

B. Obtaining phases from heavy-atom data

Consider a single reflection of amplitude $|\mathbf{F_P}|$ ($_P$ for protein) in the native data, and the corresponding reflection of amplitude $|\mathbf{F_{HP}}|$ ($_{HP}$ for heavy atom plus protein) in data from a heavy-atom derivative. Because the diffractive contributions of all atoms to a reflection are additive, the difference in amplitudes $(|\mathbf{F_{HP}}| - |\mathbf{F_P}|)$ is the amplitude contribution of the heavy atom alone, and the square of this difference, $(|\mathbf{F_{HP}}| - |\mathbf{F_P}|)^2$, is proportional to the difference $I_{HP} - I_P$. (Remember that $|\mathbf{F}|$ is proportional to $I^{1/2}$.) If we compute a diffraction pattern in which the amplitude of each reflection is $(|\mathbf{F_{HP}}| - |\mathbf{F_P}|)^2$, the result is the diffraction pattern of the heavy atom alone in the protein's unit cell. In effect, we have subtracted away all contributions from the protein atoms, leaving only the heavy-atom contributions. Now we see the diffraction pattern of one (or only a small number) of atoms, rather than the far more complex pattern of the protein.

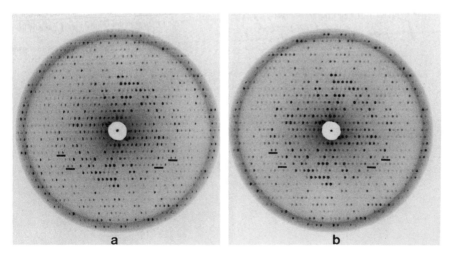

Figure 6.5 Precession photographs of the *hk*0 plane in native (*a*) and heavy-atom (*b*) crystals of the MoFe protein from nitrogenase. Corresponding underlined pairs in the native and heavy-atom patterns show reversed relative intensities. Photos courtesy of Professor Jeffrey Bolin.

In comparison to the protein structure, this "structure"— a sphere (or very few spheres) in a lattice — is very simple. It us usually easy to "determine" this structure, that is, to find the location of the heavy atom in the unit cell. Before considering how to locate the heavy atom (Section III.C), I will show how finding it helps us to solve the phase problem.

Suppose we are able to locate a heavy atom in the unit cell of derivative crystals. Recall that Equation (5.15) gives us the means to calculate the structure factors F_{hkl} for a known structure. This calculation gives us not just the amplitudes but the complete structure factors, including each of their phases. So we can compute the amplitudes and phases of our simple structure, the heavy atom in the protein unit cell. Now consider a single reflection hkl as it appears in the native and derivative data. Let the structure factor of the native reflection be F_P. Let the structure factor of the corresponding derivative reflection be F_{HP}. Finally, let F_H be the structure factor for the heavy atom itself, which we can compute if we can locate the heavy atom.

Figure 6.6 shows the relationship among the vectors F_P, F_{HP}, and F_H on the complex plane. (Remember that we are considering this relationship for a specific reflection, but the same relationship holds for all reflections.) Because the diffractive contributions of atoms are additive vectors,

$$F_{HP} = F_H + F_P \tag{6.8}$$

That is, the structure factor for the heavy-atom derivative is the vector sum of the structure factors for the protein alone and the heavy atom alone.

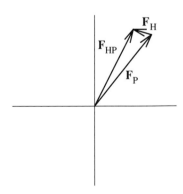

Figure 6.6 A structure factor F_{HP} for the heavy-atom derivative is the sum of contributions from the native structure (F_P) and the heavy atom (F_H).

For each reflection, we wish to know $\mathbf{F_P}$. (We already know that its length is obtainable from the measured reflection intensity I_P, but we want to learn its phase angle.) According to the previous equation,

$$\mathbf{F_P} = \mathbf{F_{HP}} - \mathbf{F_H} \tag{6.9}$$

We can solve this vector equation for $\mathbf{F_P}$, and thus obtain the phase angle of the structure factor, by representing the equation in the complex plane (Fig. 6.7).

We know $|\mathbf{F_{HP}}|$ and $|\mathbf{F_P}|$ from measuring reflection intensities I_{HP} and I_P. So we know the length of the vectors $\mathbf{F_{HP}}$ and $\mathbf{F_P}$, but not their directions or phase angles. We know $\mathbf{F_H}$, including its phase angle, from locating the heavy atom and calculating all its structure factors. To solve Equation (6.9) for $\mathbf{F_H}$ and thus obtain its phase angle, we place the vector $-\mathbf{F_H}$ at the origin and draw a circle of radius $|\mathbf{F_{HP}}|$ centered on the head of vector $-\mathbf{F_H}$ (Fig. 6.7a). All points on this circle equal the vector sum $|\mathbf{F_{HP}}| - \mathbf{F_H}$. In other words, we know that the head of $\mathbf{F_{HP}}$ lies somewhere on this circle of radius $|\mathbf{F_{HP}}|$. Next we add a circle of radius $|\mathbf{F_P}|$ centered at the origin (Fig. 6.7b). We know that the head of the vector $\mathbf{F_P}$ lies somewhere on this circle, but we do not know where because we do not know its phase angle. Equation (6.9) holds only at points where the two circles intersect. Thus the phase angles of the two vectors $\mathbf{F_P^a}$ and $\mathbf{F_P^b}$ that terminate at the points of intersection of the circles are the only possible phases for this reflection.

Our heavy-atom derivative allows us to determine, for each reflection hkl, that α_{hkl} has one of two values. How do we decide which of the two

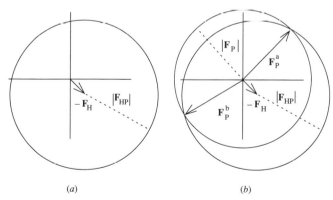

(a)   (b)

Figure 6.7 Vector solution of Equation (6.9). (a) All points on the circle equal the vector sum $|\mathbf{F_{HP}}| - \mathbf{F_H}$. (b) Vectors from the origin to intersections of the two circles are solutions to Equation (6.9).

phases is correct? In some cases, if the two intersections lie near each other, the average of the two phase angles will serve as a reasonable estimate. I will show in Chapter 7 that certain phase improvement methods can sometimes succeed with such phases from only one derivative, in which case the structure is said to be solved by the method of *single isomorphous replacement* (SIR). More commonly, however, a second heavy-atom derivative must be found and the vector problem outlined above must be solved again. Of the two possible phase angles found by using the second derivative, one should agree better with one of the two solutions from the first derivative, as shown in Fig. 6.8.

Figure 6.8a shows the phase determination using a second heavy-atom derivative; F'_H is the structure factor for the second heavy atom. The radius of the smaller circle is $|F'_{HP}|$, the amplitude of F'_{HP} for the second heavy-atom derivative. For this derivative, $F_P = F'_{HP} - F'_H$. Construction as before shows that the phase angles of F_P^c and F_P^d are possible phases for this reflection. In Fig. 6.8b, the circles from Figs. 6.7b and 6.8a are superimposed, showing that F_P^c is identical to F_P^a. This common solution to the two vector equations is F_P, the desired structure factor. The phase of this reflection is therefore the angle labeled α in the figure, the only phase compatible with data from both derivatives.

In order to resolve the phase ambiguity from the first heavy-atom derivative, the second heavy atom must bind at a different site from the first. If two heavy atoms bind at the same site, the phases of F_H will be the same in both cases, and both phase determinations will provide the same information. This is true because the phase of an atomic structure factor depends only on the location of the atom in the unit cell and not on its identity

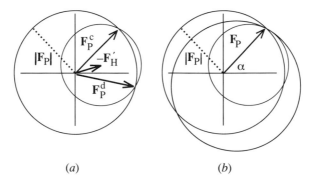

(a)   (b)

Figure 6.8 (a) A second heavy-atom derivative indicates two possible phases, one of which corresponds to F_a in Fig. 6.7b. (b) F_P, which points from the origin to the common intersection of the three circles, is the solution to Equation (6.9) for both heavy-atom derivatives. Thus α is the correct phase for this reflection.

(Chapter 5, Section III.A). In practice, it sometimes takes three or more heavy-atom derivatives to produce enough phase estimates to make the needed initial dent in the phase problem. Obtaining phases with two or more derivatives is called the method of *multiple isomorphous replacement* (MIR). This is the method by which most protein structures have been determined.

To compute a high-resolution structure, we must ultimately know the phases α_{hkl} for all reflections. High-speed computers can solve large numbers of these vector problems rapidly, yielding an estimate of each phase along with a measure of its precision.[1] For many phases, the precision of the first phase estimate is so low that the phase is unusable. For instance, in Fig. 6.9, the circles graze each other rather than intersecting sharply, so there is a large uncertainty in α. In some cases, because of inevitable experimental errors in measuring intensities, the circles do not intersect at all.

Computer programs for calculating phases also compute statistical parameters representing attempts to judge the quality of phases. Some parameters, usually called *phase probabilities*, are measures of the uncertainty of individual phases. Others parameters, including figure of merit, closure errors, phase differences, and various R-factors are attempts to assess the quality of groups of phases obtained by averaging results from several heavy-atom derivatives (or results from other phasing methods). In most cases, these parameters are numbers between 0 (poor phases) and 1 (perfect phases). No single one of these statistics is an accurate measure of the goodness of phases. Crystallographers often use two or more of these criteria simultaneously in order to cull out questionable phases. In short, until

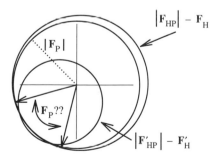

Figure 6.9 The MIR solution for this structure factor gives phase of high uncertainty.

[1] Computer programs calculate phases for a each derivative numerically (rather than geometrically) by obtaining two solutions to the equation $F_{HP}^2 = F_P^2 + F_H^2 + 2F_P F_H \cos(\alpha_P + \alpha_H)$. The pairs of solutions for heavy-atom derivatives should have one solution in common.

correct phases are obtained (see Chapter 7), there is no sure way to measure the quality of estimates. The acid test of phases is whether they give an interpretable electron-density map.

When promising phases are available, the crystallographer carries out Fourier synthesis [Equation (6.7)] to calculate $\rho(x,y,z)$. The Fourier series at this point will contain only those structure factors with promising phase estimates. In some approaches, each Fourier term is multiplied by the associated phase probability. This procedure gives greater weight to terms with more probable phases. Every phase that defies solution or is too uncertain (and for that matter every intensity that is too weak to measure accurately) forces the crystallographer to omit one term from the Fourier series when calculating $\rho(x,y,z)$. Each omitted term lowers the accuracy of the approximation to $\rho(x,y,z)$, degrading the quality and resolution of the resulting map. In practice, a good pair of heavy-atom derivatives may allow us to estimate only a small percentage of the phases. We can enlarge our list of precise phases by iterative processes mentioned briefly in Chapter 5, Section IV.B, which I will describe more fully in Chapter 7. For now, I will complete this discussion of the heavy-atom method by considering how to find heavy atoms, which is necessary for calculation of $\mathbf{F}_H$.

C. Locating heavy atoms in the unit cell

Before we can obtain phase estimates by the method described in the previous section, we must locate the heavy atoms in the unit cell of derivative crystals. As I described above, this entails extracting the relatively simple diffraction signature of the heavy atom from the far more complicated diffraction pattern of the heavy-atom derivative, and then solving a simpler "structure," that of one heavy atom (or a few) in the unit cell of the protein. The most powerful tool in determining the heavy-atom coordinates is a Fourier series called the *Patterson function $P(u,v,w)$*, a variation of the Fourier series used to compute $\rho(x,y,z)$ from structure factors. The coordinates (u,v,w) locate a point in a *Patterson map*, in the same way that coordinates (x,y,z) locate a point in an electron-density map. The Patterson function or Patterson synthesis is a Fourier series without phases. The amplitude of each term is the square of one structure factor, which is proportional to the measured reflection intensity. Thus we can construct this series from intensity measurements, even though we have no phase information. Here is the Patterson function in general form:

$$P(u,v,w) = \frac{1}{V}\sum_h\sum_k\sum_l |F_{hkl}^2| \cdot e^{-2\pi i(hu+kv+lw)} \qquad (6.10)$$

To obtain the Patterson function solely for the heavy atoms in derivative crystals, we construct a difference Patterson function, in which the amplitudes are $(\Delta F)^2 = (|F_{HP}| - |F_P|)^2$. The difference between the structure-factor amplitudes with and without the heavy atom reflects the contribution of the heavy atom alone. The difference Patterson function is

$$\Delta P(u,v,w) = \frac{1}{V}\sum_h\sum_k\sum_l \Delta F_{hkl}^2 \cdot e^{-2\pi i(hu+kv+lw)} \qquad (6.11)$$

In words, the difference Patterson function is a Fourier series of simple sine and cosine terms. (Remember that the exponential term is shorthand for these trigonometric functions.) Each term in the series is derived from one reflection hkl in both the native and derivative data sets, and the amplitude of each term is $(|F_{HP}| - |F_P|)^2$, which is the amplitude contribution of the heavy atom to structure factor F_{HP}. Each term has three frequencies: h in the u direction, k in the v direction, and l in the w direction. Phases of the structure factors are not included; at this point, they are unknown.

Because the Patterson function contains no phases, it can be computed from any raw set of crystallographic data, but what does it tell us? A contour map of $\rho(x,y,z)$ displays areas of high density (peaks) at the locations of atoms. In contrast, a Patterson map, which is a contour map of $P(u,v,w)$, displays peaks at locations corresponding to vectors between atoms. (This is a strange idea at first, but the example given below will make it clearer.) Of course, there are more vectors between atoms than there are atoms, so a Patterson map is more complicated than an electron-density map. But if the structure is simple, like that of a few heavy atoms in the unit cell, the Patterson map may be simple enough to allow us to locate the atom(s). You can see now that the main reason for using the difference Patterson function instead of a simple Patterson using F_{HP}'s is to eliminate the enormous number of peaks representing vectors between light atoms in the protein.

I will show, in a two-dimensional example, how to construct the Patterson map from a simple crystal structure and then how to use a calculated Patterson map to deduce a structure. The simple molecular structure in Fig. 6.10a contains three atoms (dark circles) in each unit cell. To construct the Patterson map, first draw all possible vectors between atoms in one unit cell, including vectors between the same pair of atoms but in opposite directions. (For example, treat $1 \rightarrow 2$ and $2 \rightarrow 1$ as distinct vectors.) Two of

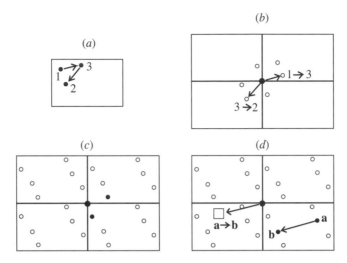

Figure 6.10 Construction and interpretation of a Patterson map. (*a*) Structure of unit cell containing three atoms. Two of the six interatomic vectors are shown. (*b*) Patterson map is constructed by moving all interatomic vectors to the origin. Patterson "atoms" (peaks in the contour map) occur at the head of each vector. (*c*) Complete Patterson map, containing all peaks from (*b*) in all unit cells. Peak at origin results from self-vectors. Image of original structure is present (origin and two darkened peaks) amid other peaks. (*d*) Trial solution of map (*c*). If origin and Patterson atoms **a** and **b** were the image of the real unit cell, the interatomic vector **a** → **b** would produce a peak in the small box. Absence of the peak disproves this trial solution.

the six vectors (1 → 3 and 3 → 2) are shown in the figure. Then draw empty unit cells around an origin (Fig. 6.10*b*), and redraw all vectors with their tails at the origin. The head of each vector is the location of a peak in the Patterson map, sometimes called a Patterson "atom" (light circles). The coordinates (u,v,w) of a Patterson atom representing a vector between atom 1 at (x_1,y_1,z_1) and atom 2 at (x_2,y_2,z_2) are $(u,v,w) = (x_1 - x_2, y_1 - y_2, z_1 - z_2)$. The vectors from Fig. 6.10*a* are redrawn in Fig. 6.10*b*, along with all additional Patterson atoms produced by this procedure. Finally, in each unit cell, duplicate the Patterson atoms from all four unit cells. The result (Fig. 6.10*c*) is a complete Patterson map of the structure in Fig. 6.10*a*. In this case, there are six Patterson atoms in each unit cell. You can easily prove to yourself that a real unit cell containing n atoms will give a Patterson unit cell containing $n(n - 1)$ Patterson atoms.

Now let's think about how to go from a computed Patterson map to a structure, that is, how to locate real atoms from Patterson atoms. A computed Patterson map exhibits a strong peak at the origin, because this is the location of all vectors between an atom and itself. Notice in Fig. 6.10c that the origin and two of the Patterson atoms (dark circles) reconstruct the original arrangement of atoms in Fig. 6.10a. Finding six peaks (ignoring the peak at the origin) in each unit cell of the calculated Patterson map, we infer that there are three real atoms per unit cell. [Solve the equation $n(n - 1) = 6$.] We therefore know that the origin and two peaks reconstruct the relationship among the three real atoms, but we do not know which two peaks to choose. To solve the problem, we pick a set of peaks (the origin and two others) as a trial solution and follow the rules described above to generate the expected Patterson map for this arrangement of atoms. If the trial map has the same peaks as the calculated map, then the trial arrangement of atoms is correct. By trial and error, we can determine which pair of Patterson atoms, along with an atom at the origin, would produce the remaining Patterson atoms. Figure 6.10d shows an incorrect solution (the origin plus peaks **a** and **b**). The vector **a** → **b** is redrawn at the origin to show that the map does not contain the Patterson atom **a** → **b**, and hence that this solution is incorrect.

You can see that as the number of real atoms increases, the number of Patterson atoms, and with it the difficulty of this problem, increases rapidly. Computer programs can search for solutions to such problems and, finding a solution, can refine the atom positions by least-squares methods (Chapter 7) to give the most likely arrangement of heavy atoms.

Unit-cell symmetry can also simplify the search for peaks in a three-dimensional Patterson map. For instance, in a unit cell with a 2_1 axis (twofold screw) on edge **c**, recall (equivalent positions, Chapter 4, Section II.H) that each atom at (x, y, z) has an identical counterpart atom at $(-x, -y, \frac{1}{2} + z)$. The vectors connecting such symmetry-related atoms will all lie at $(u, v, w) = (2x, 2y, \frac{1}{2})$ in the Patterson map (just subtract one set of coordinates from the other), which means that they all lie in the plane that cuts the Patterson unit cell at $w = \frac{1}{2}$. Such planes, which contain the Patterson vectors for symmetry-related atoms, are called *Harker sections* or *Harker planes*. If heavy atoms bind to the protein at equivalent positions, heavy-atom peaks in the Patterson map can be found on the Harker sections. (Certain symmetry elements give Patterson vectors that all lie on a line, called a *Harker line*, rather than on a plane.)

An added complication: the arrangement of heavy atoms in a protein unit cell is often enantiomeric. For example, if heavy atoms are found along a threefold screw axis, the screw may be left- or right-handed. The Patterson map does not distinguish between mirror-image arrangements of heavy atoms. But the phases obtained by calculating structure factors from the wrong enantiomer are incorrect and will not lead to an interpretable map.

Crystallographers refer to this difficulty as the "hand problem." If derivative data are available to high resolution, the crystallographer simply calculates two electron-density maps, one with phases from each enantiomer of the heavy-atom structure. With luck, one of these maps will be distinctly clearer than the other. If derivative data is available only at low resolution, this method may not determine the hand with certainty. The problem may require the use of anomalous scattering methods, discussed in Section IV.E.

Having located the heavy atom(s) in the unit cell, the crystallographer can compute the structure factors $\mathbf{F_H}$ for the heavy atoms alone, using Equation (5.15). This calculation yields both the amplitudes and the phases of structure factors $\mathbf{F_H}$, giving the vector quantities needed to solve Equation (6.9) for the phases α_{hkl} of protein structure factors $\mathbf{F_P}$. This completes the information needed to compute a first electron-density map, using Equation (6.7). This map requires improvement, because these first phase estimates contain substantial errors. I will discuss improvement of phases and maps in Chapter 7.

IV. Anomalous scattering

A. Introduction

A second means of obtaining phases from heavy-atom derivatives takes advantage of the heavy atom's capacity to absorb x-rays of specified wavelength. As a result of this absorption, Friedel's law (Chapter 4, Section III.G) does not hold, and the reflections hkl and $-h-k-l$ are not equal in intensity. This inequality of symmetry-related reflections is called *anomalous scattering* or *anomalous dispersion*.

Recall from Chapter 4, Section III.B that elements *absorb* x-rays as well as emit them, and that this absorption drops sharply at wavelengths just below their characteristic emission wavelengths (Fig. 4.16). This sudden change in absorption as a function of λ is called an *absorption edge*. An element exhibits anomalous scattering when the x-ray wavelength is near the element's absorption edge. Absorption edges for the light atoms in the unit cell are not near the wavelength of x-rays used in crystallography, so carbon, nitrogen, and oxygen do not contribute to anomalous scattering. However, absorption edges of heavy atoms are in this range, and if x-rays of varying wavelength are available, as is often the case at synchrotron sources, x-ray data can be collected under conditions that maximize anomalous scattering by the heavy atom.

B. The measurable effects of anomalous scattering

When the x-ray wavelength is near the heavy-atom absorption edge, a fraction of the radiation is absorbed by the heavy atom and reemitted with altered phase. The effect of this anomalous scattering on a given structure factor F_{HP} in the heavy-atom data is depicted in vector diagrams as consisting of two perpendicular contributions, one real (ΔF_r), the other imaginary (ΔF_i).

In Fig. 6.11, $F_{HP}^{\lambda 1}$ represents a structure factor for the heavy-atom derivative measured at wavelength λ_1, where anomalous scattering does not occur; $F_{HP}^{\lambda 2}$ is the same structure factor measured at a second x-ray wavelength λ_2 near the absorption edge of the heavy atom, so anomalous scattering alters the heavy-atom contribution to this structure factor. The vectors representing anomalous scattering contributions are ΔF_r (real) and ΔF_i (imaginary). From the diagram, you can see that

$$F_{HP}^{\lambda 2} = F_{HP}^{\lambda 1} + \Delta F_r + \Delta F_i \qquad (6.12)$$

Figure 6.12 shows the result of anomalous scattering for a Friedel pair of structure factors, distinguished from each other in the figure by superscripts + and −. Recall that for Friedel pairs in the absence of anomalous scattering, $|F_{hkl}| = |F_{-h-k-l}|$ and $\alpha_{hkl} = -\alpha_{-h-k-l}$, so $F_{HP}^{\lambda 1-}$ is the reflection of $F_{HP}^{\lambda 1+}$ in the real axis. The real contributions ΔF_r^+ and ΔF_r to the reflections of a Friedel pair are, like the structure factors themselves, reflections of each other in the real axis. On the other hand, it can be shown (but I will not prove it here) that the imaginary contribution to $F_{HP}^{\lambda 1-}$ is the inverted reflection of that for $F_{HP}^{\lambda 1+}$. That is, ΔF_i^- is obtained by reflecting ΔF_i^+ in

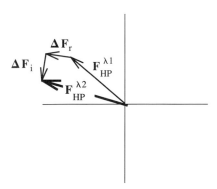

Figure 6.11 Real and imaginary anomalous-scattering contributions alter the magnitude and phase of the structure factor.

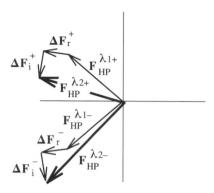

Figure 6.12 Under anomalous scattering, $\mathbf{F}_{-h-k-l}$ is no longer the mirror image of $\mathbf{F}_{hkl}$.

the real axis and then reversing its sign or pointing it in the opposite direction. Because of this difference between the imaginary contributions to these reflections, under anomalous scattering the two structure factors are no longer precisely equal in intensity, nor are they precisely opposite in phase. It is clear from Fig. 6.12 that $\mathbf{F}_{HP}^{\lambda 2-}$ is not the mirror image of $\mathbf{F}_{HP}^{\lambda 2+}$. From this disparity between Friedel pairs, the crystallographer can extract phase information.

C. Extracting phases from anomalous scattering data

The magnitude of anomalous scattering contributions $\Delta\mathbf{F}_r$ and $\Delta\mathbf{F}_i$ for a given element are constant and roughly independent of reflection angle θ, so these quantities can be looked up in tables of crystallographic information. The phases of $\Delta\mathbf{F}_r$ and $\Delta\mathbf{F}_i$ depend only on the position of the heavy atom in the unit cell, so once the heavy atom is located by Patterson methods, the phases can be computed. The resulting full knowledge of $\Delta\mathbf{F}_r$ and $\Delta\mathbf{F}_i$ allows Equation (6.12) to be solved for the vector $\mathbf{F}_{HP}^{\lambda 1}$, thus establishing its phase. Crystallographers obtain solutions by computer, but I will solve the general equation using complex vector diagrams (Fig. 6.13) and thus show that the amount of information is adequate to solve the problem.

First consider the structure factor $\mathbf{F}_{HP}^{\lambda 1+}$ in Fig. 6.12. Applying Equation (6.12) and solving for $\mathbf{F}_{HP}^{\lambda 1+}$ gives

$$\mathbf{F}_{HP}^{\lambda 1+} = \mathbf{F}_{HP}^{\lambda 2+} - \Delta\mathbf{F}_r^+ - \Delta\mathbf{F}_i^+ \qquad (6.13)$$

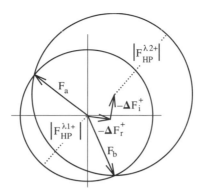

Figure 6.13 Vector solution of Equation (6.13). $\Delta\mathbf{F}_r$ and $\Delta\mathbf{F}_i$ play the same role as $\mathbf{F}_H$ in Figs. 6.7 and 6.8.

To solve this equation (see Fig. 6.13), draw the vector $-\Delta\mathbf{F}_r^+$ with its tail at the origin, and draw $-\Delta\mathbf{F}_i^+$ with its tail on the head of $-\Delta\mathbf{F}_r^+$. With the head of $-\Delta\mathbf{F}_i^+$ as center, draw a circle of radius $|\mathbf{F}_{HP}^{\lambda 2+}|$, representing the amplitude of this reflection in the anomalous scattering data set. The head of the vector $\mathbf{F}_{HP}^{\lambda 2+}$ lies somewhere on this circle. We do not know where, because we do not know the phase of the reflection. Now draw a circle of radius $|\mathbf{F}_{HP}^{\lambda 1+}|$ with its center at the origin, representing the structure-factor amplitude of this same reflection in the nonanomalous scattering data set. The two points of intersection of these circles satisfy Equation (6.13), establishing the phase of this reflection as either that of $\mathbf{F}_a$ or $\mathbf{F}_b$. As with the SIR method, we cannot tell which of the two phases is correct.

The Friedel partner of this reflection comes to the rescue. We can obtain a second vector equation involving $\mathbf{F}_{HP}^{\lambda 1+}$ by reflecting $\mathbf{F}_{HP}^{\lambda 2-}$ and all its vector components across the real axis (Fig. 6.14a).

After reflection, $\mathbf{F}_{HP}^{\lambda 1-}$ equals $\mathbf{F}_{HP}^{\lambda 1+}$, $\Delta\mathbf{F}_r^-$ equals $\Delta\mathbf{F}_r^+$, and $\Delta\mathbf{F}_i^-$ equals $-\Delta\mathbf{F}_i^+$. The magnitude of $\mathbf{F}_{HP}^{\lambda 2-}$ is unaltered by reflection across the real axis. If we make these substitutions in Equation (6.13), we obtain

$$\mathbf{F}_{HP}^{\lambda 1+} = |\mathbf{F}_{HP}^{\lambda 2-}| - \Delta\mathbf{F}_r^+ + \Delta\mathbf{F}_i^+ \tag{6.14}$$

We can solve this equation in the same manner as we solved Equation (6.13), by placing the vectors $-\Delta\mathbf{F}_r^+$ and $+\Delta\mathbf{F}_i^+$ head-to-tail at the origin, and drawing a circle of radius $|\mathbf{F}_{HP}^{\lambda 2-}|$ centered on the head of $+\Delta\mathbf{F}_i^+$ (Fig. 6.14b). Finally, we draw a circle of radius $|\mathbf{F}_{HP}^{\lambda 1+}|$ centered at the origin. The circles intersect at the two solutions to Equation (6.14). Although the circles graze each other and give two phases with considerable uncertainty,

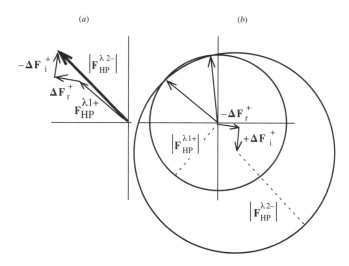

Figure 6.14 Reflection of $\mathbf{F}^-$ components across the real axis gives a second vector equation involving the desired structure factor. (*a*) All reflected components are labeled with their equivalent contributions from $\mathbf{F}^+$. (*b*) Vector solutions of Equation (6.14). These solutions are compatible only with $\mathbf{F}_a$ in Fig. 6.13.

one of the possible solutions corresponds to $\mathbf{F}_a$ in Fig. 6.13, and neither of them is close to the phase of $\mathbf{F}_b$.

So the disparity between intensities of Friedel pairs in the anomalous scattering data set establishes their phases in the nonanomalous scattering data set. The reflection whose phase has been established here corresponds to the vector $\mathbf{F}_{HP}$ in Equation (6.9). Thus the amplitudes and phases of two of the three vectors in the Equation (6.9) are known: (1) $\mathbf{F}_{HP}$ is known from the anomalous scattering computation just shown, and (2) $\mathbf{F}_H$ is known from calculating the heavy-atom structure factors after locating the heavy atom by Patterson methods. The vector $\mathbf{F}_P$, then, is simply the vector difference $\mathbf{F}_{HP} - \mathbf{F}_H$, establishing the phase of this reflection in the native data.

D. Summary

Under anomalous scattering, the members of a Friedel pair can be used to establish the phase of a reflection in the heavy-atom derivative data, thus establishing the phase of the corresponding reflection in the native data. Let me review briefly the entire project of obtaining the initial structure factors by SIR with anomalous scattering. First, we collect a complete data set with native crystals, giving us the amplitudes $|\mathbf{F}_{Phkl}|$ for each native reflection. Then we find a heavy-atom derivative and collect a second data set at the same wavelength, giving amplitudes $|\mathbf{F}_{HPhkl}|$ for each reflection in the heavy-atom data. Next we collect a third data set at a different x-ray wavelength, chosen to maximize anomalous scattering by the heavy atom. We use the nonequivalence of Friedel pairs in the anomalous scattering data to establish phases of reflections in the heavy-atom data, and we use the phased heavy-atom derivative structure factors to establish the native phases. (Puff, puff!)

In practice, several of the most commonly used heavy atoms (including uranium, mercury, and platinum) give strong anomalous scattering with $Cu-K_\alpha$ radiation. In such cases, crystallographers can measure intensities of Friedel pairs in the heavy-atom data set. In phase determination (refer to Figs. 6.12–6.14), the average of $|\mathbf{F}_{hkl}|$ and $|\mathbf{F}_{-h-k-l}|$ serves as both $|\mathbf{F}_{HP}^{\lambda1+}|$ and $|\mathbf{F}_{HP}^{\lambda1-}|$, while $|\mathbf{F}_{hkl}|$ and $|\mathbf{F}_{-h-k-l}|$ separately serve as $|\mathbf{F}_{HP}^{\lambda2+}|$ and $|\mathbf{F}_{HP}^{\lambda2-}|$, so only one heavy-atom data set is required.

Like phases from the MIR method, anomalous scattering phases can serve as only an initial estimate and must be weighted with some measure of phase probability. The intensity differences between Friedel pairs are very small, so measured intensities must be very accurate if any usable phase information is to be derived. To improve accuracy, the crystallographer collects intensities of Friedel partners under very similar conditions, and always from the same crystal. Diffractometry is ideal for anomalous scattering because of its inherently greater accuracy in measuring intensities, and because the diffractometer can be programmed to collect Friedel pairs in succession, thus ensuring that the crystal is in the same condition during collection of the two reflections.

E. Anomalous scattering and the hand problem

As I discussed in Section III.C, Patterson methods do not allow us to distinguish between enantiomeric arrangements of heavy atoms, and phases derived from heavy-atom positions of the wrong hand are incorrect. When high-resolution data are available for the heavy-atom derivative, phases and electron-density maps can be calculated for both enantiomeric possibilities. The map calculated with phases from the correct enantiomer will sometimes be demonstrably sharper and more interpretable. If not, and if anomalous scattering data are available, SIR *and* anomalous scattering phases can be computed for both hands, and maps prepared from the two sets of phases. The added phase information from anomalous scattering sometimes makes hand selection possible when SIR phases alone do not.

The availability of two heavy-atom derivatives, one with anomalous scattering, allows a powerful technique for establishing the hand, even at quite low resolution. Heavy atoms in the first derivative are located by Patterson methods, one of the possible hands is chosen, and SIR phases are computed. Then, using the same hand assumption, we can compute anomalous scattering phases. For the second heavy-atom derivative, instead of using Patterson methods, we compute a difference Fourier between the native data and the second derivative data, using the SIR phases from the first derivative. Then we compute a second difference Fourier, adding the phases from anomalous scattering. Finally, we compute a third difference Fourier, just like the second except that the signs of all anomalous scattering contributions are reversed, which is like assuming the opposite hand. The first Fourier should exhibit electron-density peaks at the positions of the second heavy atom. If the initial hand assumption were correct, heavy atom peaks should be stronger in the second Fourier. If it were incorrect, heavy atom peaks should be stronger in the third Fourier.

F. The phase problem for smaller molecules

Methods involving heavy atoms apply almost exclusively to large molecules (500 or more atoms, not counting hydrogens). For small molecules (up to 200 atoms), phases can be determined by what are commonly called *direct methods*, which rely on the existence of mathematical relationships among certain combinations of phases. From these relationships, enough initial phase estimates can be obtained to begin converging toward a complete set of phases.

Direct methods work when the number of reflections is relatively small. Isomorphous replacement works when the molecule is large enough that a heavy atom does not disturb its structure significantly. At the moment, the most difficult structures for crystallographers are those of 200–500 atoms, which are too large for direct methods and too small to remain isomorphous despite the intrusion of a heavy atom. Newer NMR methods are of great power for medium-size molecules. If a medium-size molecule naturally contains a heavier atom, like iron or zinc, it can sometimes be phased by anomalous scattering. The following method applies to all molecules, regardless of size, but requires knowledge that the desired structure is similar to a known structure.

V. Molecular replacement: Related proteins as phasing models

A. Introduction

The crystallographer can sometimes use the phases from structure factors of a known protein as initial estimates of phases for a new protein. If this method is feasible, the crystallographer may be able to determine the structure of the new protein from a single native data set. The known protein in this case is referred to as a *phasing model*, and the method, which entails calculating initial phases by placing a model of the known protein in the unit cell of the new protein, is called *molecular replacement*.

For instance, the mammalian serine proteases, trypsin, chymotrypsin, and elastase, are very similar in structure and conformation. If a new mammalian serine protease is discovered, and sequence homology with known proteases suggest that this new protease is similar in structure to known ones, one of the known proteases might be used as a phasing model for determining the structure of the new protein.

Similarly, having learned the crystallographic structure of a protein, we may want to study the conformational changes that occur when the protein binds to a small ligand, and to learn the molecular details of protein–ligand binding. We might be able to crystallize the protein and ligand together, or introduce the ligand into protein crystals by soaking. We expect that the protein/ligand complex is similar in structure to the free protein. If this expectation is realized, we do not have to work completely from scratch to determine the structure of the complex. We can use the ligand-free protein as a phasing model for the protein/ligand complex.

B. Isomorphous phasing models

If the phasing model and the new protein are isomorphous, as may be the case when a small ligand is soaked into protein crystals, then the phases from the free protein can be used directly to compute $\rho(x,y,z)$ from native intensities of the new protein [Equation (6.15)].

$$\rho\,(x,\,y,\,z)\;=\;\frac{1}{V}\sum_{h}\sum_{k}\sum_{l}\left|F_{hkl}^{\mathrm{new}}\right|e^{\,-2\pi i\,(hx\,+\,ky\,+\,lz\,-\,\alpha_{hkl}^{\prime\,\mathrm{model}})}\qquad\qquad(6.15)$$

In this Fourier synthesis, the amplitudes $\left|F_{hkl}^{\mathrm{new}}\right|$ are obtained from the native intensities of the new protein, and the phases α'^{model} are those of the phasing model. During the iterative process of phase improvement (Chapter 7), the phases should change from those of the model to those of the new protein or complex, revealing the desired structure.

C. Nonisomorphous phasing models

If the phasing model is not isomorphous with the desired structure, the problem is more difficult. The phases of atomic structure factors, and hence of molecular structure factors, depend on the location of atoms in the unit cell. In order to use a known protein as a phasing model, we must superimpose the structure of the model on the structure of the new protein in its unit cell and then calculate phases for the properly oriented model. In other words, we must find the position and orientation of the phasing model in the new unit cell that would give phases most like those of the new protein. Then we can calculate the structure factors of a properly positioned model and use the phases of these computed structure factors as initial estimates of the desired phases.

Without knowing the structure of the new protein, how can we copy the model into the unit cell with the proper orientation and position? From native data on the new protein, we can determine its unit-cell dimensions and symmetry. Clearly the phasing model must be placed in the unit cell with the same symmetry as the desired protein. This places some constraints on where to place the model, but not enough to give useful estimates of phases. In theory, it should be possible to conduct a computer search of all orientations and positions of the model in the new unit cell. For each trial position and orientation, we would calculate the structure factors (called $\mathbf{F}_{\mathrm{calc}}$) of the model [Equation (5.15)] and compare their amplitudes $\left|\mathbf{F}_{\mathrm{calc}}\right|$ with the measured amplitudes $\left|\mathbf{F}_{\mathrm{obs}}\right|$ obtained from diffraction intensities of the new protein. Finding the position and orientation that gives the best match, we would take the computed phases (α_{calc}) as the starting phases for structure determination of the new protein.

D. Separate searches for orientation and location

In practice, the number of trial orientations and positions for the phasing model is enormous, so a brute-force search is impractical, even on the fastest computers. The procedure is greatly simplified by separating the search for the best orientation from the search for the best position. Further, it is possible to search for the best orientation independently of location by using the Patterson function.

If you consider the procedure for drawing a Patterson map from a known structure (Section III.C), you will see that the final map is independent of the position of the structure in the unit cell. No matter where you draw the "molecule," as long as you do not change its orientation (that is, as long as you do not rotate it within the unit cell), the Patterson map looks the same. On the other hand, if you rotate the structure in the unit cell, the Patterson map rotates around the origin, altering the arrangement of Patterson atoms in a single Patterson unit cell. This suggests that the Patterson map might provide a means of determining the best orientation of the model in the unit cell of the new protein.

If the model and the new protein are indeed similar, and if they are oriented in the same way in unit cells of the same dimensions and symmetry, they should give very similar Patterson maps. We might imagine a trial and-error method in which we compute Patterson maps for various model orientations and compare them with the Patterson map of the desired protein. In this manner, we could find the best orientation of the model, and then use that single orientation in our search for the best position of the model, using the structure-factor approach outlined above.

How much computing do we actually save by searching for orientation and location separately? The orientation of the model can be specified by three angles of rotation about orthogonal axes $x, y,$ and z with their origins at the center of the model. Specifying location also requires three numbers, the $x, y,$ and z coordinates of the molecular center with respect to the origin of the unit cell. For sake of argument, let us say that we must try 100 different values for each of the six parameters. (In real situations, the number of trial values is much larger.) The number of combinations of six parameters, each with 100 possible values, is 100^6, or 10^{12}. Finding the orientation as a separate search requires first trying 100 different values for each of three angles, which is 100^3 or 10^6 combinations. After finding the orientation, finding the location requires trying 100 different values of each of three coordinates, again 100^3 or 10^6 combinations. The total number of trials for separate orientation and location searches is $10^6 + 10^6$ or 2×10^6. The magnitude of the saving is $10^{12}/(2 \times 10^6)$ or 500,000. In this case, the problem of finding the orientation and location separately is smaller by half a million times than the problem of searching for orientation and location simultaneously.

E. Monitoring the search

Finally, what mathematical criteria are used in these searches? In other words, as the computer goes through sets of trial values (angles or coordinates) for the model, how does it compare results and determine optimum values of the parameters?

For the orientation search (often called a "rotation search"), the computer is looking for large values of the model Patterson function $P^{model}(u,v,w)$ at locations corresponding to peaks in the Patterson map of the desired protein. A powerful and sensitive way to evaluate the model Patterson is to compute the minimum value of $P^{model}(u,v,w)$ at all locations of peaks in the Patterson map of the desired protein. A value of zero for this minimum means that the trial orientation has no peak in at least one location where the desired protein exhibits a peak. A high value for this minimum means that the trial orientation has peaks at all locations of peaks in the Patterson map of the desired protein.

For the location search, the criterion is the correspondence between the expected structure-factor amplitudes from the model in a given trial location and the actual amplitudes derived from the native data on the desired protein. This criterion can be expressed as the R-factor, a parameter we will encounter later as a criterion of improvement of phases in final structure determination. The R-factor compares overall agreement between the amplitudes of two sets of structure factors, as follows:

$$R = \frac{\sum ||\mathbf{F}_{obs}| - |\mathbf{F}_{calc}||}{\sum |\mathbf{F}_{obs}|} \qquad (6.16)$$

In words, for each reflection, we compute the difference between the observed structure-factor amplitude from the native data set $|\mathbf{F}_{obs}|$ and the calculated amplitude from the model in its current trial location $|\mathbf{F}_{calc}|$, and take the absolute value, giving the magnitude of the difference. We add these magnitudes for all reflections. Then we divide by the sum of the observed structure-factor amplitudes.

If, on the whole, the observed and calculated amplitudes agree with each other, the differences in the numerator are small, and the sum of the differences is small compared to the sum of the amplitudes themselves, so R is small. For perfect agreement, all the differences equal zero and R equals zero. No single difference is likely to be larger than the corresponding $|\mathbf{F}_{obs}|$, so the maximum value of R is one. For proteins, R values of 0.3–0.4 for the best placement of a phasing model have often provided adequate initial estimates of phases.

F. Summary

If we know that the structure of a new protein is similar to that of a known protein, we can use the known protein as a phasing model and thus solve the phase problem without heavy-atom derivatives. If the new crystals and those of the model are isomorphous, the model phases can be used directly as estimates of the desired phases. If not, we must somehow superimpose the known protein on the new protein to create the best phasing model. We can do this without knowledge of the structure of the new protein by using Patterson-map comparisons to find the best orientation of the model protein and then using structure-factor comparisons to find the best location of the model protein.

VI. Iterative improvement of phases (preview of Chapter 7)

The phase problem greatly increases the effort required to obtain an interpretable electron-density map. In this chapter, I have discussed several methods of obtaining phases. In all cases, the phases obtained are estimates, and often the set of estimates is incomplete. Electron-density maps calculated from Equation (6.7), using measured amplitudes and first phase estimates, are often difficult or impossible to interpret. In Chapter 7, I will discuss improvement of phase estimates and extension of phase assignments to as many reflections as possible. As phase improvement and extension proceed, electron-density maps become clearer and easier to interpret as an image of a molecular model. The iterative process of structure refinement eventually leads to a structure that is in good agreement with the original data.

7 Obtaining and Judging the Molecular Model

I. Introduction

In this chapter, I will discuss the final stages of structure determination: obtaining and improving the electron-density map, interpreting the map to produce an atomic model of the unit-cell contents, and refining the model to optimize its agreement with the original native reflection intensities. The criteria by which the crystallographer judges the progress of the work overlap with criteria for assessing the quality of the final model. These subjects form the bridge from Chapter 7 to Chapter 8, where I will review many of the concepts of this book by guiding you through the experimental descriptions from a recent structure determination.

II. Iterative improvement of maps and models: Overview

In brief, obtaining a detailed molecular model of the unit-cell contents entails calculating $\rho(x,y,z)$ from Equation (6.7) using measured amplitudes from the native data set and phases computed from heavy-atom data, anomalous scattering, or molecular replacement. Because the phases are rough estimates, the first map may be uninformative and disappointing. Crystallographers improve the map by an iterative process sometimes called "bootstrapping." The basic principle of this iteration is easy to state but demands care, judgment, and much labor to execute: any features that can be reliably discerned in the map become part of a phasing model for subsequent maps.

Whatever crude model of unit-cell contents that can be discerned in the map is cast in the form of a simple electron-density function and used to calculate new structure factors by Equation (5.16). The phases of these structure factors are used, along with the original native amplitudes, to add more terms to Equation (6.7), the Fourier-series description of $\rho(x,y,z)$, in hopes of producing a clearer map. When the map becomes clear enough to allow location of atoms, structure factors are computed using Equation (5.15), which contains atomic structure factors rather than electron density. As the model becomes more detailed, the phases computed from it improve, and the model, computed from the original native structure-factor amplitudes and the latest phases, becomes even more detailed. The crystallographer thus tries to bootstrap from the initial rough phase estimates to phases of high accuracy, and from them, to a clear, interpretable map and a model that fits the map well.

The model can be improved in another way: by least-squares refinement of the atomic coordinates. This method entails adjusting the atomic coordinates to improve the agreement between amplitudes calculated from the current model and the original measured amplitudes in the native data set. In the latter stages of structure determination, the crystallographer alternates between map interpretation and least-squares refinement.

The block diagram in Fig. 7.1 shows how these various methods ultimately produce a molecular model that agrees with the native data. The vertical dotted line in Fig. 7.1 divides the operations into two categories. To the right of the the line are real-space methods, which entail attempts to improve the electron-density map, by adding information to the map or removing noise from it, or to improve the model, using the map as a guide. To the left of the line are reciprocal-space methods, which entail attempts to improve phases or improve the agreement between reflection intensities computed from the model and the original measured reflection intensities. In real-space methods, the criteria for improvement or removal of errors are found in electron-density maps, in the fit of model to map, or in the adherence of the model to expected bond lengths and angles (all real-space criteria); in reciprocal-space methods, the criteria for improvement or removal of errors involve reliability of phases and agreement of calculated structure

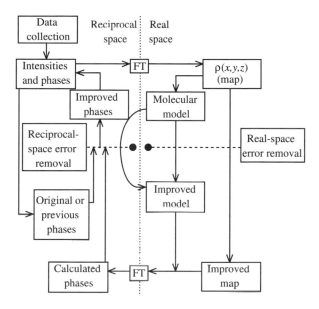

Figure 7.1 Block diagram of crystallographic structure determination.

factors with measured intensities (all reciprocal-space criteria). The link between real and reciprocal space is, of course, the Fourier transform (FT).

I will return to this diagram near the end of the chapter, particularly to amplify the meaning of the term *error removal*, which is indicated by dashed horizontal lines in Fig. 7.1. For now, I will illustrate the bootstrapping technique for improving phases, map, and model with an analogy: the method of successive approximations for solving a complicated algebraic equation. Most mathematics education emphasizes equations that can be solved analytically for specific variables. Many realistic problems defy such analytic solutions, but are amenable to numerical methods. The method of successive approximations has much in common with the iterative process that extracts a protein model from diffraction data.

Consider the problem of solving the following equation for the variable y.

$$\left(1 + \frac{1}{y^2}\right) \cdot (y - 1) = 1 \tag{7.1}$$

Attempts to simplify the equation produce a cubic equation in y, giving no straightforward means to an analytical solution. You can, however, easily obtain a numerical solution for y with a hand calculator. Start by solving for y in terms of y^2 as follows:

$$y = \cfrac{1}{\left(1 + \cfrac{1}{y^2}\right)} + 1 \tag{7.2}$$

Then make an arbitrary initial estimate of y, say, $y = 1$. (This is analogous to starting with the MIR phases as initial estimates of the correct phases.) Plug this estimate into the right-hand y^2 term, and calculate y [analogous to computing a crude structure from measured structure-factor amplitudes ($|\mathbf{F}_{obs}|$) and phase estimates]. The result is 1.5. Now take this computed result as the next estimate (analogous to computing new structure factors from the crude structure), plug it into the y^2 term, and compute y again (analogous to computing a new structure from better phase estimates). The result is 1.6923. Repeating this process produces these answers in succession: 1.7412, 1.752, 1.7543, 1.7547, 1.7549, 1.7549, and so on. After a few iterations, the process converges to a solution; that is, the output value of y is the same as the input. This value is a solution to the original equation.

With Equation (7.2) above, any first estimate above 1.0 (even one million) produces the result shown. In contrast, for many other equations, the method of successive approximations works only if the initial estimate is close to a correct solution. Otherwise, the successive answers do not converge; instead, they may oscillate among several values (the iteration "hangs up" instead of converging) or may continually become larger in magnitude (the iteration "blows up"). In order for the far more complex crystallographic iteration to converge to a protein model that is consistent with the diffraction data, initial estimates of many phases must be close to the correct values. Attempts to start from random phases in hopes of convergence to correct ones appear to be doomed to failure because of the large number of incorrect solutions to which the process can converge.

The following sections describe the crystallographic bootstrapping process in more detail.

III. First maps

A. Resources for the first map

Entering the final stages of structure determination, the crystallographer is armed with several sets of data with which to calculate $\rho(x,y,z)$ as a Fourier series of structure factors using Equation (5.18). First is the original native data set, which usually contains the most accurate and complete (highest resolution) set of measured intensities. These data will support the most critical tests of the final molecular model. Next are data sets from heavy-atom derivatives, which are often limited to lower resolution. Several sets of phases may be available, calculated from heavy-atom derivatives and perhaps from anomalous dispersion. Because each phase must be calculated from a heavy-atom reflection, phase estimates are not available for native reflections at resolutions higher than that of the best heavy-atom derivative. Finally, for each set of phases, there is usually some criterion of precision. These criteria will be used as weighting factors, numbers between 0 and 1, for Fourier terms containing the phases. A Fourier term containing a phase estimate of low reliability will be multiplied by a low weighting factor in the Fourier-series computation of $\rho(x,y,z)$. In other words, such a term will be multiplied by a number less than 1.0 to reduce its contribution to the Fourier series, and thus reduce bias from a reflection whose phase is questionable. Conversely, a term containing a phase of high reliability will be given full weight (weighting factor of 1.0) in the series.

Here is the Fourier series that gives the first electron-density map:

$$\rho\,(x,y,z) \;=\; \frac{1}{V}\sum_{h}\sum_{k}\sum_{l}\; w_{hkl}|\mathbf{F}_{obs}|\,e^{-2\pi i(hx+ky+lz-\alpha'_{calc})} \tag{7.3}$$

In words, the desired electron density function is a Fourier series in which term hkl has amplitude $|\mathbf{F}_{obs}|$, which equals $(I_{hkl})^{1/2}$, the square root of the measured intensity I_{hkl} from the native data set. The phase α'_{hkl} of the same term is calculated from heavy-atom, anomalous dispersion, or molecular replacement data, as described in Chapter 6. The term is weighted by the factor w_{hkl}, which will be near 1.0 if α'_{hkl} is among the most highly reliable phases, or smaller if the phase is questionable. This Fourier series is called an $\mathbf{F}_{obs}$ or $\mathbf{F}_o$ synthesis (and the map an $\mathbf{F}_o$ map) because the amplitude of each term hkl is $|\mathbf{F}_{obs}|$ for reflection hkl.

The first term in this Fourier series, the $\mathbf{F}_{000}$ term, should contain $(I_{000})^{1/2}$, where I_{000} is the intensity of reflection 000, which lies at the origin of the reciprocal lattice. Recall that this reflection is never measured because it is obscured by the direct beam. Examination of Equation (7.3) reveals that $\mathbf{F}_{000}$ is a real constant (as opposed to a complex or imaginary number). The phase α'_{000} of this term is assigned a value of zero, with the result that all other phases will be computed relative to this assignment. Then because $h = k = l = 0$ for reflection 000, the exponent of e is zero and the entire exponential term is 1.0. Thus $\mathbf{F}_{000}$ is a constant, just like $\mathbf{f}_0$ in Fig. 2.14.

All other terms in the series are simple trigonometric functions with average values of zero, so it is clear that the value assigned to F_{000} will determine the overall amplitude of the electron-density map. (In the same manner, the f_0 term in Fig. 2.14 displaces all the Fourier sums upward, making the sums positive for all values of x, like the target function.) The sensible assignment for F_{000} is therefore the total number of electrons in the unit cell, making the sum of $\rho(x,y,z)$ over the whole unit cell equal to the total electron density. In practice, this term can be omitted from the calculation, and the overall map amplitude can be set by means described in Section III.C, below.

B. Displaying and examining the map

The contour map of the first calculated electron density is often displayed by printing sections of the unit cell onto Plexiglas or clear plastic sheets and stacking them to produce a three-dimensional model, called a *minimap*. This first glimpse of the molecular image is often attended with great excitement and anticipation. If the phase estimates are sufficiently good, the minimap will show some of the gross features of unit-cell contents. In the rare best cases, with good phases from molecular replacement, and perhaps with enhancement from noncrystallographic averaging, first maps are easily interpretable, clearly showing continuous chains of electron density and features like alpha helices, perhaps even allowing some amino-acid side chains to be identified. At the worst, the first map is singularly uninformative, signaling the need for additional phasing information, perhaps from another heavy-atom derivative. Usually the minimum result that promises a structure from the existing data is that protein be distinguishable from bulk water. If the bounds of each molecule, the molecular envelope, can be defined in the first map, then a structure is likely to come forth.

I will consider the latter case, in which the first map defines a molecular envelope, but little additional detail is visible. If more detail can be discerned, the crystallographer can jump ahead to later stages of the map-improvement process I am about to describe. If the molecular envelope cannot be discerned, more data collection is required.

C. Improving the map

The crude molecular image seen in the F_o map, which is obtained from the original indexed intensity data ($|F_{obs}|$) and the first phase estimates (α'_{calc}), serves now as a model of the desired structure. A crude electron-density function is devised to describe the unit-cell contents as well as they can be observed in the first map. Then the function is modified to make it more realistic in the light of known properties of proteins and water in crystals. This process is called, depending on the exact details of procedure, *density modification, solvent leveling,* or *solvent flattening.*

The electron-density function devised by density modification may be no more than a fixed, high value of $\rho(x,y,z)$ for all regions that appear to be within a protein molecule, and a fixed, low value of ρ for all surrounding areas of bulk solvent. One automated method first defines the molecular envelope by dividing the unit cell into a grid of regularly spaced points. At each point, the value of $\rho(x,y,z)$ in the $\mathbf{F}_o$ map is evaluated. At each grid point, if ρ is negative, it is reassigned a value of zero; if ρ is positive, it is assigned a value equal to the average value of ρ within a defined distance of the grid point. This procedure smooths the map; eliminates many small, random fluctuations in density; and essentially divides the map into two types of regions: those of relatively high (protein) and relatively low (solvent) density. Next, the overall amplitude of the map is increased until the ratio of high density to low density agrees with the ratio of protein to solvent in the crystal, either assuming that the crystal is about half water, or using a value derived from the measured crystal density (Chapter 3, Section IV). This contrived function $\rho(x,y,z)$ is now used to compute structure factors, using Equation (5.16). From this computation, we learn what the amplitudes and phases of all reflections would be if this new model were correct. We use the phases from this computation, which constitute a new set of α'_{hkl}'s, along with the $|\mathbf{F}_{obs}|$'s derived from the original measured intensities, to calculate $\rho(x,y,z)$ again, using Equation (7.3).

We do not throw out old phases immediately but continue to weight each Fourier term with some measure of phase quality. In this manner, we continue to let the data speak for itself as much as possible, rather than allowing the current model to bias the results. If the new phase estimates are better, then the new $\rho(x,y,z)$ will be improved, and the electron-density map will be more detailed. The new map serves to define the molecular boundary more precisely, and the cycle is repeated. (Refer again to the block diagram in Fig. 7.1.) If we continue to use good judgment in incorporating new phases and new terms into Equation (7.3), successive Fourier-series computations of $\rho(x,y,z)$ include more terms, and successive contour maps become clearer and more interpretable. In other words, the iterative process of incorporating phases from successively better and more complete models converges toward a structure that fits the native data better. The phase estimates "converge" in the sense that the output phases computed from the current model [Equation (5.16)] agree better with the input phases that went into computation of the model [Equation (7.3)].

As this process continues, and the model becomes more detailed, we begin to get estimates for the phases of structure factors at resolution beyond that of the heavy-atom derivatives. In a process called *phase extension*, we gradually increase the number of terms in the Fourier series of Equation (7.3), adding terms that contain native intensities (as $|\mathbf{F}_{obs}|$) at slightly higher resolution with phases from the current model. This must be done gradually and judiciously, so as not to let incorrect areas of the current model bias the calculations excessively. If the new phase estimates are good, the resulting map has slightly higher resolution, and structure factors computed from Equation (5.16) give useful phase estimates at still higher resolution. In this manner, low-resolution phases are improved, and phase assignments are extended to higher resolution.

If phase extension seems like getting something from nothing, realize that by using general knowledge about protein and solvent density, we impose justifiable restrictions on the model, giving it realistic properties that are not visible in the map. In effect, we are using known crystal properties to increase the resolution of the model. Thus it is not surprising that the phases calculated from the modified model are good to higher resolution than those calculated from an electron-density function that does little more than describe what can be seen in the map.

Another means of improving the map at this stage depends on the presence of noncrystallographic symmetry elements in the unit cell. Recall that the intensity of reflections results from many molecules in identical orientations diffracting identically. In a sense, the diffraction pattern is the sum of diffraction patterns from all individual molecules. This is equivalent to taking a large number of weak, noisy signals (each the diffraction from one molecule) and adding them together to produce a strong signal. The noise in the individual signals, which might include the background intensity of the film or the weak signal of stray x-rays, is random, and when many weak signals are added, this random noise cancels out.

In some cases, the strength of this signal can be increased further by averaging the signals from molecules that are identical, but have different orientations in the unit cell, such that no two orientations of the crystal give the same orientation of these molecules in the x-ray beam. These molecules may be related by symmetry elements that are not aligned with symmetry elements of the entire unit cell. Thus the diffractive contributions of these identical molecules are never added together. In such cases, the unit cell is said to exhibit noncrystallographic symmetry. By knowing the arrangement of molecules in the unit cell, that is, by knowing the location and type of noncrystallographic symmetry elements, the crystallographer can use a computer to simulate the movement of these sets of molecules into identical orientations, and thus add their signals together. The result is improved signal-to-noise ratio, and in the end, a clearer image of the molecules. This method, called *symmetry averaging*, is spectacularly successful in systems with a high degree of symmetry, such as viruses. Many virus-coat proteins are icosahedral, possessing two-, three-, and fivefold rotation axes. Often one or more two- and threefold axes are noncrystallographic, and fivefold axes are always noncrystallographic, because no unit cell exhibits fivefold symmetry.

IV. The model becomes molecular

A. New phases from the molecular model

At some critical point in the iterative improvement of phases, the map becomes clear enough that we can trace the protein chain through it. For instance, we may be able to recognize alpha helices, one of the densest features of a protein, or sheets of beta structure. Now we can construct a partial molecular model (as opposed to an electron-density model) of the protein, using computer graphics to build and manipulate a stick model of the known sequence within small sections of the map (called *map fitting,* discussed below). From this model, which may harbor many errors and undefined regions, we again calculate structure factors, this time using Equation (5.15), which treats each atom in the current model as an independent scatterer. In other words, we calculate new structure factors from our current molecular model rather than from an approximation of $\rho(x,y,z)$. Additional iterations may improve the map further, allowing more features to be constructed therein.

Here again, as in density modification, we are using known properties of proteins to improve the model beyond what we can actually see in the map. Thus we are in effect improving the resolution of the model by making it structurally realistic: giving it local electron densities corresponding to the light atoms that we know are present, and connecting atoms at bond lengths and angles that we know must be correct. So again, our successive models give us phases for reflections at higher and higher resolution. Electron-density maps computed from these phases, and, as always, the original native amplitudes $|\mathbf{F}_{obs}|$, become more and more detailed.

B. Minimizing bias from the model

Conversion to a molecular model greatly increases the hazard of introducing excessive bias from the model into $\rho(x,y,z)$. At this point, bias can be decreased by one of several alternative Fourier computations of the electron-density map. As phases from the model begin to be the most reliable, they begin to dominate the Fourier series. In the extreme, the series would contain amplitudes purely from the intensity data and phases purely from the model. In order to compensate for the increased influence of model phases, and to continue letting the intensity data influence improvement of the model, the crystallographer calculates electron-density maps using various difference Fourier syntheses, in which the amplitude of each term is of the form $(|n|\mathbf{F}_{obs}| - |\mathbf{F}_{calc}|)$ which reduces overall model influence by subtracting the calculated structure-factor amplitudes $(|\mathbf{F}_{calc}|)$ from some multiple of the observed amplitudes $(|\mathbf{F}_{obs}|)$ within each Fourier term. For $n = 1$, the Fourier series is called an $\mathbf{F}_o - \mathbf{F}_c$ synthesis:

$$\rho\,(x,y,z) \;=\; \frac{1}{V}\sum_h \sum_k \sum_l \; \big||\mathbf{F}_o| - |\mathbf{F}_c|\big|\, e^{-2\pi i(hx + ky + lz - \alpha'_{calc})} \tag{7.4}$$

A contour map of this Fourier series is called an "$\mathbf{F}_o - \mathbf{F}_c$" map. How is this map interpreted? Depending on which of $|\mathbf{F}_o|$ or $|\mathbf{F}_c|$ is larger, Fourier terms can be either positive or negative. The resulting electron-density map contains both positive and negative "density." Positive density in a region of the map implies that the contribution of the observed intensities ($|\mathbf{F}_o|$'s) to ρ are larger than the contribution of the model ($|\mathbf{F}_c|$'s), and thus that the unit cell (represented by $|\mathbf{F}_o|$'s) contains more electron density in this region than implied by the model (represented by $|\mathbf{F}_c|$'s). In other words, the map is telling us that the model should be adjusted to increase the electron density in this region, by moving atoms toward the region. On the other hand, a region of negative density indicates that the model implies more electron density in the region than the unit cell actually contains. The region of negative density is telling us to move atoms away from this region. As an example, if an amino-acid side chain in the model is in the wrong conformation, the $\mathbf{F}_o - \mathbf{F}_c$ map may exhibit a negative peak coincident with the erroneous model side chain and a nearby positive peak signifying the correct position.

The $\mathbf{F}_o - \mathbf{F}_c$ map emphasizes errors in the current model, but it lacks the familiar appearance of the molecular surface found in an $\mathbf{F}_o$ map. In addition, if the model still contains many errors, the $\mathbf{F}_o - \mathbf{F}_c$ map is "noisy," full of small positive and negative peaks that are difficult to interpret. The

$F_o - F_c$ map is most useful near the end of the structure determination, when most of the model errors have been eliminated. The $F_o - F_c$ map is a great aid in detecting subtle errors after most of the serious errors are corrected.

A more easily interpreted and intuitively satisfying difference map, but one that still allows undue influence by the model to be detected, is the $2F_o - F_c$ map, calculated as follows:

$$\rho\,(x,y,z)\ =\ \frac{1}{V}\sum_h\sum_k\sum_l\ |2|F_o|-|F_c||\,e^{-2\pi i(hx+ky+lz-\alpha'_{calc})} \tag{7.5}$$

In this map, the model influence is reduced, but not as severely as with $F_o - F_c$. Unless the model contains extremely serious errors, this map is everywhere positive, and contours at carefully chosen electron densities resemble a molecular surface. With experience, the crystallographer can often see the bias of an incorrect area of the model superimposed on the true signal of the correct structure as implied by the original intensity data. For instance, in a well-refined map (see model refinement below), backbone carbonyl oxygens are found under a distinct bulge in the backbone electron density. If a carbonyl oxygen in the model is pointing 180° away from the actual position in the molecule, the bulge in the map may be weaker than usual, or misshapen (sometimes cylindrical), and a weak bulge may be visible on the opposite side of the carbonyl carbon, at the true oxygen position. Correcting the oxygen orientation in the model, and then recalculating structure factors, results in loss of the weak, incorrect bulge in the map, and intensification of the bulge in the correct location. (This may sound like a serious correction of the model, requiring the movement of many atoms, but the entire peptide bond can be flipped 180° around the backbone axis with only slight changes in the positions of neighboring atoms.)

Various other Fourier syntheses are used during these stages in order to improve the model. Some crystallographers prefer a $3F_o - 2F_c$ map, a compromise between $F_o - F_c$ and $2F_o - F_c$, for the final interpretation. In areas where the maps continue to be ambiguous, it is often helpful to examine the original MIR or molecular replacement maps for insight into how model building in this area might be started off on a different foot. Another measure is to eliminate the atoms in the questionable region and calculate structure factors from Equation (5.15), so that the possible errors in the region contribute nothing to the phases, and hence do not bias the resulting map, which is called an "omit map" or "chop map." (Another important type of difference Fourier synthesis, which is used to compare similar protein structures, is discussed in Chapter 8, Section III.C.)

C. Map fitting

Conversion to a molecular model is usually done piecemeal, as the map reveals recognizable structural features. This procedure, called *map fitting* or *model building*, entails interpreting the electron density map by building a molecular model that fits realistically into the molecular surface implied by the map. In modern crystallographic labs, map fitting is done by interactive computer graphics. A computer program produces a realistic three-dimensional display of small sections of one or more electron-density maps, and allows the user to construct and manipulate molecular models to fit the map. The viewer sees the model within the map, as shown in Plate 2b. As the model is constructed or adjusted, the program stores current atom locations in the form of three-dimensional coordinates. The crystallographer, while building a model interactively on the computer screen, is actually building a list of atoms, each with a set of coordinates (x, y, z) to specify its location. Coordinates are automatically updated whenever the model is adjusted. This list of coordinates is the output file from the map-fitting program and the input file for calculation of new structure factors. When the model is correct and complete, this file becomes the means by which the model is shared with the community of scientists who study proteins (see Section VII of this chapter).

In addition to routine commands for inserting or changing amino-acid residues, moving atoms and fragments, and changing conformations, map-fitting programs contain many sophisticated tools to aid the model builder. Fragments, treated as rigid assemblies of atoms, can be automatically fitted to the map by the method of least squares (see Section V.A). After manual adjustments of the model, which may result in unrealistic bond lengths and angles, portions of the model can be "regularized," which entails automatic correction of bond lengths and angles with minimal movement of atoms. In effect, regularization looks for the most realistic configuration of the model that is very similar to its current configuration. Where small segments of the known sequence cannot be easily fitted to the map, some map-fitting programs can search the Protein Data Bank (see Section VII) for fragments having the same sequence, and then display these fragments so that the user can see whether they fit the map.

Following is a somewhat idealized description of how map fitting may proceed, illustrated with views from a modern map-fitting program. The maps and models are from the structure determination of adipocyte lipid-binding protein (ALBP), which I will discuss further in Chapter 8.

When the map has been improved to the point that molecular features are revealed, the crystallographer attempts to trace the protein through as much continuous density as possible. At this point the quality of the map will vary from place to place, perhaps being quite clear in the molecular interior, which is usually more ordered, and exhibiting broken density in some places, particularly at chain termini and surface loops. Because we know that amino-acid side chains branch regularly off α-carbons in the main chain, we can estimate the positions of many α-carbons. These atoms should lie near the center of the main-chain density next to bulges that represent side chains. In proteins, α-carbons are 3.8–4.2 Å apart. This knowledge allows the crystallographer to construct an α-carbon model of the molecule (Plate 8), and to compute structure factors from this model.

Further improvement of the map with these phases may reveal side chains more clearly. Now the trick is to identify some specific side chains so that the known amino-acid sequence of the protein can be aligned with visible features in the map. As mentioned above, chain termini are often ill-defined, so we need a foothold for alignment of sequence with map where the map is sharp. Often the key is a short stretch of sequence containing several bulky hydrophobic residues, like Trp, Phe, and Tyr (tryptophan, phenylalanine, and tyrosine). Because they are hydrophobic, they are likely to be in the interior where the map is clearer. Because they are bulky, their side-chain density is more likely to be identifiable. From such a foothold, the detailed model building can begin.

Regions that cannot be aligned with sequence are often built with poly-alanine, reflecting our knowledge that all amino acids contain the same backbone atoms, and all but one, glycine, have at least a β-carbon (Plate 9). In this manner, we build as many atoms into the model as possible in the face of our ignorance about how to align the sequence with the map in certain areas.

In pleated sheets, we know that successive carbonyl oxygens point in opposite directions. One or two carbonyls whose orientations are clearly revealed by the map can allow sensible guesses as to the positions of others within the same sheet. As mentioned above, in map fitting, we use knowledge of protein structure to infer more than the map shows us. If our inferences are correct, subsequent maps, computed with phases calculated from the model, will show enhanced evidence for the inferred features and will show additional features as well, leading to further improvement of the model. Poor inferences degrade the map; so where electron density conflicts with intuition, we follow the density as closely as possible.

With each successive map, new molecular features are added as they can be discerned, and errors in the model, such as side-chain conformations that no longer fit the electron density, are corrected. As the structure nears completion, the crystallographer may use $2F_o - F_c$ and $F_o - F_c$ maps simultaneously to track down the most subtle disagreements between the model and the data.

V. Structure refinement

A. Least-squares methods

Cycles of map calculation and model building, which are forms of real-space refinement of the model, are interspersed with computerized attempts to improve the agreement of the model with the original intensity data. (Everything goes back to those original reflection intensities, which give us our $|\mathbf{F}_{obs}|$ values!) Because these computations entail comparison of computed and observed structure factor amplitudes (reciprocal space), rather than examination of maps and models (real space), these methods are referred to as *reciprocal-space refinement*. Most commonly, this process is a massive version of least-squares fitting, the same procedure that freshman chemistry students employ to construct a straight line that fits a scatter graph of data.

In the simple least-squares method in two dimensions, the aim is to find a function $y = f(x)$ that fits a series of observations (x_1, y_1), (x_2, y_2), . . . , (x_i, y_i), where each observation is a data point, a measured value of the independent variable x at some selected value y. (For example, y might be the temperature of a gas and x might be its measured pressure.) The solution to the problem is a function $f(x)$ for which the sum of the squares of distances between the data points and the function itself is as small as possible. In other words, $f(x)$ is the function that minimizes D, the sum of the squared differences between observed (y_i) and calculated $[f(x_i)]$ values, as follows:

$$D = \sum_i w_i \, (y_i - f(x_i))^2 \tag{7.6}$$

The differences are squared to make them all positive; otherwise, for a large number of random differences, D simply equals zero. The term w_i is an optional weighting factor that reflects the reliability of observation i, thus giving greater influence to the most reliable data. According to principles of statistics, w_i should be $1/(\sigma_i)^2$, where σ_i is the standard deviation computed from multiple measurements of the same data point (x_i, y_i).

In the simplest case, $f(x)$ is a straight line, for which the general equation is $f(x) = mx + b$, where m is the slope of the line and b is the intercept of the line on the $f(x)$ axis. Solving this problem entails finding the proper values of the parameters m and b. If we substitute $(mx_i + b)$ for each $f(x_i)$ in Equation (7.6), take the partial derivative of the right-hand side with respect to m and set it equal to zero, and then take the partial derivative with respect to b and set it equal to zero, the result is a set of simultaneous equations in m and b. Because all the squared differences are to be minimized simultaneously, the number of equations equals the number of observations, and there must be at least two observations to fix values for the two parameters m and b. With just two observations (x_1, y_1) and (x_2, y_2), m and b are determined precisely, and $f(x)$ is the equation of the straight line between (x_1, y_1) and (x_2, y_2). If there are more than two observations, the problem is "overdetermined" and the values of m and b describe the straight line of best fit to all the observations. So the solution to this simple least-squares problem is a pair of parameters m and b for which the function $f(x) = mx + b$ minimizes D.

B. Crystallographic refinement

In the crystallographic case, the parameters we seek (analogous to m and b) are, for all atoms j, the positions (x_j, y_j, z_j) that best fit the observed structure-factor amplitudes. Because the positions of atoms in the current model can be used to calculate structure factors, and hence to compute the *expected* structure-factor amplitudes ($|\mathbf{F}_{calc}|$) for the current model, we want to find a set of atom positions that give $|\mathbf{F}_{calc}|$'s, analogous to calculated values $f(x_i)$, that are as close as possible to the $|\mathbf{F}_{obs}|$'s (analogous to observed values y_i). In least-squares terminology, we want to select atom positions that minimize the squares of differences between corresponding $|\mathbf{F}_{calc}|$'s and $|\mathbf{F}_{obs}|$'s. We define the difference between the observed amplitude $|\mathbf{F}_{obs}|$ and the measured amplitude $|\mathbf{F}_{calc}|$ for reflection hkl as $(|\mathbf{F}_o| - |\mathbf{F}_c|)_{hkl}$, and we seek to minimize the function Φ, where:

$$\Phi = \sum_{hkl} w_{hkl} \left(|\mathbf{F}_o| - |\mathbf{F}_c| \right)^2_{hkl} \tag{7.7}$$

In words, the function Φ is the sum of the squares of differences between observed and calculated amplitudes. The sum is taken over all reflections hkl currently in use. Each difference is weighted by the term w_{hkl}, a number that depends on the reliability of the corresponding measured intensity. As in the simple example, according to principles of statistics, the weight should be $1/(\sigma_{hkl})^2$, where σ is the standard deviation from multiple measurements of $|\mathbf{F}_{obs}|$. Because the data do not usually contain enough measurements of each reflection to determine its standard deviation, other weighting schemes have been devised. Starting from a reasonable model, the least-squares refinement method succeeds about equally well with a variety of weighting systems, so I will not discuss them further.

C. Additional refinement parameters

We seek a set of parameters that minimize the function Φ. These parameters include the atom positions, of course, because the atom positions in the model determine each $\mathbf{F}_{calc}$. But other parameters are included as well. One is the temperature factor B_j of each atom j, a measure of how much the atom oscillates around the position specified in the model. Atoms at side-chain termini are expected to exhibit more freedom of movement than main-chain atoms, and this movement amounts to spreading the atoms' centers over a small region of space. Diffraction is affected by this variation in atomic position, so it is realistic to assign a temperature factor to each atom and include the factor among parameters to vary in minimizing Φ. From the temperature factors computed during refinement, we learn which atoms in the molecule have the most freedom of movement, and we gain some insight into the dynamics of our largely static model. In addition, adding the effects of motion to our model makes it more realistic and hence more likely to fit the data precisely.

Another parameter included in refinement is the occupancy n_j of each atom j, a measure of the fraction of molecules in which atom j actually occupies the position specified in the model. If all molecules in the crystal are precisely identical, then occupancies for all atoms are 1.00. Occupancy is included among refinement parameters because occasionally two or more distinct conformations are observed for a small region like a surface side chain. The model might refine better if atoms in this region are assigned occupancies equal to the fraction of side chains in each conformation. For example, if the two conformations occur with equal frequency, then atoms involved receive occupancies of 0.5 in each of their two possible positions. By including occupancies among the refinement parameters, we obtain estimates of the frequency of alternative conformations, giving some additional information about the dynamics of the protein molecule.

The factor $|\mathbf{F}_c|$ in Equation (7.7) can be expanded to show all the parameters included in refinement, as follows:

$$\mathbf{F}_c = G \cdot \sum_j n_j f_j e^{2\pi i(hx_j + ky_j + lz_j)} \cdot e^{-B_j[(\sin\theta)/\lambda]^2} \qquad (7.8)$$

Although this equation is rather forbidding, it is actually a familiar one [Equation (5.15)] with the new parameters included. Equation (7.8) says that structure factor $\mathbf{F}_{hkl}$ can be calculated ($\mathbf{F}_c$) as a Fourier series containing one term for each atom j in the current model. The term G is an overall scale factor to put all $\mathbf{F}_c$'s on a convenient numerical scale. In the jth term, which describes the diffractive contribution of atom j to this particular structure factor, n_j is the occupancy of atom j; f_j is its scattering factor, just as in Equation (5.16); x_j, y_j, and z_j are its coordinates; and B_j is its temperature factor. The first exponential term is the familiar Fourier description of a simple three-dimensional wave with frequencies h, k, and l in the directions x, y, and z. The second exponential shows that the effect of B_j on the structure factor depends on the angle of the reflection ($[\sin \theta]/\lambda$).

D. Local minima and radius of convergence

As you can imagine, finding parameters (atomic coordinates, occupancies, and temperature factors for all atoms in the model) to minimize the differences between all the observed and calculated structure factors is a massive computing task. As in the simple example, one way to solve this problem is to differentiate Φ with respect to all the parameters, which gives simultaneous equations with the parameters as unknowns. The number of equations equals the number of observations, in this case the number of measured reflection intensities in the native data set. The parameters are overdetermined only if the number of measured reflections is greater than the number of parameters to be obtained. The complexity of the equations rules out analytical solutions, and requires iterative (successive-approximation) methods that we hope will converge from the starting parameters of our current model to a set of new parameters corresponding to a minimum in Φ. It has been proved that the atom positions that minimize Φ are the same as those found from Equation (7.3), the Fourier-series description of electron density. So real-space and reciprocal-space methods converge to the same solution.

The complicated function Φ undoubtedly exhibits many *local minima*, corresponding to variations in model conformation that minimize Φ with respect to other quite similar ("neighboring") conformations. A least-squares procedure will find the minimum that is nearest the starting point, so it is important that the starting model parameters be near the global minimum, the one conformation that gives best agreement with the native structure factors. Otherwise the refinement will converge into an incorrect local minimum from which it cannot extract itself. The greatest distance from the global minimum from which refinement will converge properly is called the *radius of convergence*. The theoretically derived radius is $d_{\min}/4$, where $d_{\min}$ is the lattice-plane spacing corresponding to the reflection of highest resolution used in the refinement. Inclusion of data from higher resolution, while potentially giving more information, decreases the radius of convergence, so the model must be increasingly close to its global minimum as more data are included in refinement.

There are a number of approaches to increasing the radius of convergence and thus increasing the probability of finding the global minimum. These approaches take the form of additional constraints and restraints on the model during refinement computations. A *constraint* is a fixed value for a certain parameter. For example, in early stages of refinement, we might constrain all occupancies to a value of 1.0. A *restraint* is a subsidiary condition imposed on the parameters, such as the condition that all bond lengths and bond angles be within a specified range of values. The function Φ, with additional restraints on bond lengths and angles, is as follows:

$$
\begin{aligned}
\Phi = \;& \sum_{hkl} w_{hkl} \left(|\mathbf{F}_o| - |\mathbf{F}_c| \right)^2_{hkl} \\[2mm]
& + \sum_i^{\text{bonds}} w_i \left(d_i^{\text{ideal}} - d_i^{\text{model}} \right)^2 \\[2mm]
& + \sum_j^{\text{angles}} w_j \left(\phi_j^{\text{ideal}} - \phi_j^{\text{model}} \right)^2
\end{aligned}
\tag{7.9}
$$

where d_i is the length of bond i and ϕ_j is the bond angle at location j. Ideal values are average values for bond lengths and angles in small organic molecules, and model values are taken from the current model. In minimizing this more complicated Φ, we are seeking atom positions, temperature factors, and occupancies that simultaneously minimize differences between (1) observed and calculated reflection amplitudes, (2) model bond lengths and ideal bond lengths, and (3) model bond angles and ideal bond angles. In effect, the restraints penalize adjustments to parameters if the adjustments make the model less realistic.

E. Molecular energy and motion in refinement

In recent years, crystallographers have begun to take advantage of the prodigious power of supercomputers to include knowledge of molecular energy and molecular motion in the refinement. In energy refinement, least-squares restraints are placed on the overall energy of the model, including bond, angle, and conformational energies and the energies of noncovalent interactions such as hydrogen bonds. Adding these restraints is an attempt to find the structure of lowest energy in the neighborhood of the current model. In effect, these restraints penalize adjustments to parameters if the adjustments increase the calculated energy of the model.

Another form of refinement employs molecular dynamics, which is an attempt to simulate the movement of molecules by solving Newton's laws of motion for atoms moving within force fields that represent the effects of covalent and noncovalent bonding. Molecular dynamics can be turned into a tool for crystallographic refinement by including an energy term that is related to the difference between the measured reflection intensities and the intensities calculated from the model. In effect, this approach treats the model as if its energy decreases as its fit to the native crystallographic data improves. In refinement by simulated annealing, the model is allowed to move as if at high temperature, in hopes of lifting it out of local energy minima. Then the model is cooled slowly to find its preferred conformation at the temperature of diffraction data collection. All the while, the computer is searching for the conformation of lowest energy, with the assigned energy partially dependent on agreement with diffraction data. In some cases, the radius of convergence is greatly increased by this process, a form of molecular dynamics refinement.

VI. Convergence to a final structure

A. Producing the final map and model

In the last stages of structure determination, the crystallographer alternates computed, reciprocal-space refinement with map fitting, or real-space refinement. In general, constraints and restraints are lifted as refinement proceeds, so that agreement with the original reflection intensities is gradually given highest priority. When ordered water becomes discernible in the map, water molecules are added to the model, and occupancies are no longer constrained, to reflect the fact that a particular water site may be occupied in only a fraction of unit cells. Early in refinement, all temperature factors are assigned a starting value. Later, the value is held the same for all atoms or for groups of similar atoms (like all backbone atoms as one group, and all side-chain atoms as a separate group), but the overall value is not constrained. Finally, individual atomic temperature factors are allowed to refine independently. Early in refinement, the whole model is held rigid, to refine its position in the unit cell. Then blocks of the model are held rigid while their positions refine with respect to each other. In the end, individual atoms are freed to refine independently. This gradual release of the model to refine against the original data is an attempt to prevent it from getting stuck in local minima. Choosing when to relax specific constraints and restraints is perhaps more art than science.

Near the end of refinement, the $F_o - F_c$ map becomes rather empty except in problem areas. Map fitting becomes a matter of searching for and correcting errors in the model, which amounts to extricating the model from local minima in the reciprocal-space refinement. Wherever model atoms lie outside $2F_o - F_c$ contours, the $F_o - F_c$ map will often show the atoms within negative contours, with nearby positive contours pointing to correct locations for these atoms. Many crystalline proteins possess disordered regions, where the maps do not clear up and become unambiguously interpretable. Such regions of structural uncertainty are mentioned in published papers on the structure, and in the header information of Protein Data Bank files (see Section VII).

At the end of successful refinement, the $2F_o - F_c$ map almost looks like a space-filling model of the protein. (Refer to Plate 2b, which is the final model built into the same region shown in Plates 8 and 9). The backbone electron density is continuous, and peptide carbonyl oxygens are clearly marked by bulges in the backbone density. Side-chain density, especially in the interior, is sharp and fits the model snugly. Branched side chains, like those of valine, exhibit distinct lobes of density representing the two branches. Rings of histidine, phenylalanine, tyrosine, and tryptophan are flat, and in models of the highest resolution, aromatic rings show a clear depression or hole in the density at their centers. Looking at the final model in the final map, you can easily underestimate the difficulty of interpreting the early maps, in which backbone density is frequently weak and broken, and side chains are missing or shapeless.

You can get a rough idea of how refinement gradually reveals features of the molecule by comparing electron-density maps at low, medium, and high resolution, as in Plate 7. Each photo in this set shows a section of the final ALBP model in a map calculated with the final phases, but with $|F_{obs}|$'s limited to specified resolution. In (a), only $|F_{obs}|$'s of reflections at resolution 6 Å or greater are used. With this limit on the data (which amounts to including in the $2F_o - F_c$ Fourier series only those reflections whose indices *hkl* correspond to sets of planes with spacing d_{hkl} of 6 Å or greater), the map of this pleated-sheet region of the protein is no more than a featureless sandwich of electron density. As we extend the Fourier series to include reflections out to 4.5 Å (b), the map shows distinct, but not always continuous, tubes of density for each chain. Extending the resolution to 3.0 Å, we see density that defines the final model reasonably well, including bulges for carbonyl oxygens (red) and for side chains. Finally, at 1.6 Å, the map fits the model like a glove, zigzagging precisely in unison with the backbone of the model, and showing well-defined lobes for individual side-chain atoms.

Look again at the block diagram of Fig. 7.1, which gives an overview of structure determination. Now I can be more specific about the criteria for error removal or filtering, which is shown in the diagram as horizontal dashed lines in real and reciprocal space. Real-space filtering of the *map* entails removing noise or adding density information, as in solvent flattening. Reciprocal-space filtering of *phases* entails using only the strongest reflections (for which phases are more accurate) to compute the early maps, and using figures of merit and phase probabilities to select the most reliable phases at each stage. The molecular *model* can be filtered in either real or reciprocal space. Errors are removed in real space by improving the fit of model to map, and by allowing only realistic bond lengths and angles when adjusting the model (regularization). Here the criteria are structural parameters and congruence to the map (real space). Model errors are removed in reciprocal space (curved arrow in center) by least-squares refinement, which entails adjusting atom positions in order to bring calculated intensities into agreement with measured intensities. Here the criteria are comparative structure-factor amplitudes (reciprocal space). Using the Fourier transform, the crystallographer moves back and forth between real and reciprocal space to nurse the model into congruence with the data.

B. Guides to convergence

Judging convergence and assessing model quality are overlapping tasks. I will discuss criteria of convergence here. In Chapter 8, I will discuss some of the criteria further, particularly as they relate to the quality and usefulness of the final model.

The progress of iterative real- and reciprocal-space refinement is monitored by comparing the measured structure-factor amplitudes $|F_{obs}|$ [which are proportional to $(I_{obs})^{1/2}$] with amplitudes $|F_{calc}|$ from the current model. In calculating the new phases at each stage, we learn what intensities our current model, if correct, would yield. As we converge to the correct structure, the measured **F**'s and the calculated **F**'s should also converge. The primary measure of convergence is the *residual index*, or *R*-factor (Chapter 6, Section V.E).

$$R = \frac{\sum ||\mathbf{F}_{obs}| - |\mathbf{F}_{calc}||}{\sum |\mathbf{F}_{obs}|} \tag{7.10}$$

In this expression, each $|\mathbf{F}_{obs}|$ is derived from a measured reflection intensity and each $|\mathbf{F}_{calc}|$ is the amplitude of the corresponding structure factor calculated from the current model. Values of *R* range from zero, for perfect agreement of calculated and observed intensities, to about 0.6, the *R*-factor obtained when a set of measured amplitudes is compared with a set of random amplitudes. An *R*-factor greater than 0.5 implies that agreement between observed and calculated intensities is very poor, and many models with *R* = 0.5 or greater will not respond to attempts at improvement unless more data are available. An early model with *R* near 0.4 is promising and is likely to improve with the various refinement methods I have presented. A desirable target *R*-factor for a protein model refined with data to 2.5 Å is 0.2. Very rarely, small, well-ordered proteins may refine to *R* = 0.1, while small organic molecules commonly refine to better than *R* = 0.05.

In addition to monitoring *R* as an indicator of convergence, the crystallographer monitors various structural parameters that indicate whether the model is chemically, stereochemically, and conformationally reasonable. In a *chemically* reasonable model, the bond lengths and bond angles fall near the expected values for simple organic molecules. The usual criteria applied are the root-mean-square (rms) deviations of all the model's bond lengths and angles from an accepted set of values. A well refined model exhibits rms deviations of no more than 0.02 Å for bond lengths and 4° for bond angles.

A *stereochemically* reasonable model has no inverted centers of chirality (for instance, no D-amino acids). A *conformationally* reasonable model meets several criteria: (1) peptide bonds are nearly planar, and nonproline peptides are *trans,* except where obvious local conformational constraints produce an occasional *cis*-proline; (2) the backbone conformational angles Φ and Ψ fall in allowed ranges, as judged from Ramachandran plots of these angles (see Chapter 8); and finally, (3) torsional angles at single bonds in side chains lie within a few degrees of stable, staggered conformations. During the progress of refinement, all of these structural parameters should continually improve.

VII. Sharing the model

An intensely interested audience awaits the crystallographer's final molecular model. This audience includes researchers studying the same molecule by other methods, such as spectroscopy or kinetics, or studying metabolic pathways or diseases in which the molecule is involved. The model may serve as a basis for understanding the properties of the protein and its behavior in biological systems. It may also serve as a guide to the design of inhibitors or to engineering efforts to modify its function by methods of molecular biology.

Most crystallographers appear to believe that it is part and parcel of their work to make molecular structures available to the larger community of scientists. This belief is reflected in policies of many journals and funding organizations that require public availability of the structure as a condition of publication or financial support.

Crystallographers share the fruits of their work in the form of lists of atomic coordinates, which can be used to display and study the molecule with molecular graphics programs (Chapter 9). Less commonly, because fewer people have the resources to use them, crystallographers share the final structure factors, from which electron-density maps can be computed. Among the audience for structure factors are other crystallographers developing new techniques of data handling, refinement, or map interpretation.

On request, many authors of published crystallographic structures provide coordinate lists by computer mail directly to interested parties. I obtained coordinates of the Zif268/DNA complex shown in Plate 1 and the coordinates and maps of ALBP in this manner. But the great majority of structures are available through the Protein Data Bank (PDB) at Brookhaven National Laboratory.[1] Crystallographers can satisfy publication and funding requirements for availability of their structures by depositing coordinates with this data bank.

---

[1] The Protein Data Bank is described fully in F. C. Bernstein, T. F. Koetzle, G. J. B. Williams, E. F. Meyer, Jr., M. D. Brice, J. R. Rodgers, O. Kennard, T. Shimanouchi, and M. Tasumi, "The Protein Data Bank: A computer-based archival file for macromolecular structures," *Journal of Molecular Biology* **112**, 535–542 (1977), and E. E. Abola, F. C. Bernstein, S. H. Bryant, T. F. Koetzle, and J. Weng, "Protein Data Bank," in *Crystallographic Databases—Information Content, Software Systems, Scientific Applications*, F. H. Allen, G. Bergerhoff, and R. Sievers, eds., Data Commission of the International Union of Crystallography, Bonn–Cambridge–Chester, 1987, pp. 107–132.

The Protein Data Bank checks deposited files carefully for errors and inconsistencies, and then makes them available at modest cost in a standard text [ASCII (American Standard Code For Information Interchange)] format on magnetic tapes compatible with many computers. The PDB structure files, which are called *atomic coordinate entries*, can be viewed within editor or word-processor programs. Most molecular graphics programs read PDB files directly or use them to produce their own files in binary form for rapid access during display. In addition to the coordinate list, a PDB file contains a header or opening section with information about published papers on the protein, details of experimental work that produced the structure, and other useful information. Here is a brief description of PDB file contents. The line types, given in capital letters, are printed at the left of each line in the file. The contents of the file, in order of appearance, are as follows:

- HEADER lines, containing the file name and date.
- COMPND lines, containing the name of the protein.
- SOURCE lines, giving the organism from which the protein was obtained.
- AUTHOR lines, listing the persons who placed this data in the Protein Data Bank.
- REVDAT lines, listing all revision dates for data on this protein.
- REMARK lines, containing (1) references to journal articles about the structure of this protein and (2) general information about the contents of this file.
- SEQRES lines, giving the amino-acid sequence of the protein, with amino acids specified by three-letter abbreviation.
- HET and FORMUL lines, listing the cofactors, prosthetic groups, or other nonprotein substances present in the structure.
- HELIX, SHEET, and TURN lines, listing the elements of secondary structure in the protein.
- CRYST lines, giving the unit cell dimensions and space group.
- ORIG and SCALE lines, containing instructions for computing the positions of symmetry-related molecules in the unit cell.
- ATOM lines, containing the atomic coordinates of all protein atoms, plus their structure factors and occupancies. Atoms are listed in the order given in the paragraph following this list.
- HETATM lines, which contain the same information as ATOM lines for any *nonprotein* molecules (cofactors, prosthetic groups, and solvent molecules) included in the structure and listed in HET and FORMUL lines above.
- CONECT lines, which list bonds between nonprotein atoms in the file.
- MASTER and END lines, which mark the end of the file.

After the header comes a list of model atoms in standard order. Atoms in the PDB file are named and listed according to a standard format in an all-English version of the Greek-letter conventions used by organic chemists. For each amino acid, beginning at the N-terminus, the backbone atoms are listed in the order α-nitrogen N, α-carbon CA, carbonyl carbon C, and carbonyl oxygen O, followed by the side-chain atoms, β-carbon CB, γ-carbon CG, and so forth. In branched side chains (or rings), atoms in the two branches are numbered 1 and 2 after the proper Greek letter. For example, the atoms of aspartic acid, in the order of PDB format, are N, CA, C, O, CB, CG, OE1, and OE2. The terminal atoms of the side chain are followed in the file by atom N of the next residue. There are no markers in the file to tell where one residue begins and another ends; each N marks the beginning of the next residue.

In this form, as a PDB atomic coordinate entry, a crystallographic structure becomes a matter of public record. The final model of the molecule can then fall before the eyes of anyone equipped with a computer and an appropriate molecular display program. It is natural for the consumer of these files, as well as for anyone who sees published structures in journals or textbooks, to think of the molecule as something someone has seen more or less directly. Having read this far, you know that our crystallographic vision is quite indirect. But you probably still have little intuition about possible limits to the model's usefulness. For instance, just how precise are the relative locations of atoms? How much does molecular motion alter atomic positions? For that matter, how well does the model fit the original diffraction data from which it was extracted? These and other questions are the subject of Chapter 8, in which I will start you off toward becoming a discriminating consumer of the crystallographic product. This entails understanding several criteria of model quality, and being able to extract these criteria from published accounts of crystallographic structure determination.

8 A User's Guide to Crystallographic Models

I. Introduction

Most biochemists will never determine a protein structure by x-ray crystallography. But many will at some time use a crystallographic model in research or teaching. In research, study of molecular models by computer graphics has become an indispensable tool in formulating mechanisms of protein action (for instance, binding or catalysis), searching for modes of interaction between molecules, choosing sites to modify by chemical methods or site-specific mutagenesis, and designing inhibitors of proteins involved in disease. Because protein chemists would like to learn the rules of protein folding, every new model is a potential test for proposed theories of folding, as well as for schemes for predicting conformation from amino-acid sequence. In education, modern texts in biology and chemistry are effectively and dramatically illustrated with graphics images, often as stereo pairs. Projection monitors allow instructors to show "real-time" graphics displays in the classroom, giving students vivid, animated, three-dimensional views of complex molecules.

In all of these applications, there is a tendency to treat the model as a physical entity, as a real object seen or filmed. How much confidence in the crystallographic model is justified? For instance, how precisely does crystallography establish the positions of atoms in the molecule? Are all atoms' positions equally well established? How does one rule out the possibility that crystallizing the protein alters it in some significant way? The model is a static image of a dynamic molecule, a springy system of atoms that breathes with characteristic vibrations, and tumbles dizzily through solution, as it executes its function. Does crystallography give us any insight into these motions? Are parts of the molecule more flexible than others? Are major movements of structural elements essential to the molecule's action? How does the user decide whether proposed motions of the molecule are reasonable?

In this chapter, I will discuss the strengths and limitations of molecular models obtained by x-ray diffraction. My aim is to help you use crystallographic models wisely and appropriately, and realize just what is known, and what is unknown, about a molecule that has yielded some of its secrets to crystallographic analysis. To demonstrate how you can draw these conclusions for yourself with regard to a particular molecule of interest, I will conclude this chapter by discussing a recent structure determination, as it appeared in a biochemical journal. Here my goals are (1) to help you learn to extract criteria of model quality from published structural reports, and (2) to review some basic concepts of protein crystallography.

II. Judging the quality and usefulness of the refined model

A. Structural parameters

As discussed in Chapter 7, Section VI.B, crystallographers monitor the R-factor as an indicator of convergence to a final, refined model, with a general target of 0.20 for proteins, and adequate additional cycles of refinement to confirm that R is not still declining. In addition, various constraints and restraints are relaxed during refinement, and after these restricted values are allowed to refine freely, they should remain in, or converge to, reasonable values. Among these are the root-mean-square (rms) deviations of the model's bond lengths, angles, and conformational angles from an accepted set of values based on the geometry of small organic molecules. A refined model should exhibit rms deviations of no more than 0.02 Å for bond lengths and 4° for bond angles. These values are routinely calculated during refinement to be sure that all is going well.

In effect, protein structure determination is a search for the conformation of a molecule whose chemical composition is known. For this reason, conformational angles about single bonds are not constrained during refinement, and they should settle into reasonable values. Spectroscopic evidence abundantly implies that peptide bonds are planar, and some refinements constrain peptide geometry. If unconstrained, peptide bonds should settle down to within one to two degrees of planar.

The other backbone conformational angles are Φ, along the $N-C_\alpha$ bond and Ψ, along the $C_\alpha-C$ bond, as shown in Fig. 8.1. In this figure, Φ is the torsional angle of the $N-C_\alpha$ bond, defined by the atoms $C-N-C_\alpha-C$ (C is the carbonyl carbon), and Ψ is the torsional angle of the $C_\alpha-C$ bond, defined by the atoms $N-C_\alpha-C-N$. In the figure, $\Phi = \Psi = 180°$.

Model studies show that, for each amino acid, the pair of angles Φ,Ψ is greatly restricted by steric repulsion. The allowed pairs of values are depicted on a Ramachandran diagram (Fig. 8.2). A point (Φ,Ψ) on the diagram represents the conformational angles Φ and Ψ on either side of the α-carbon of one residue. Irregular polygons enclose backbone conformational angles that do not give steric repulsion (inner polygons) or give only modest repulsion (outer polygons). Location of the letters α and β correspond to conformational angles of residues in α helix and β pleated sheet.

During the final stages of map-fitting and crystallographic refinement, Ramachandran diagrams are a great aid in finding conformationally unrealistic regions of the model. Crystallographic software packages and map-fitting programs usually contain a routine for computing Φ and Ψ for each residue from the current coordinate list, as well as for generating the

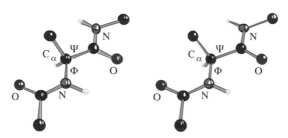

Figure 8.1 Backbone conformational angles in proteins (stereo).

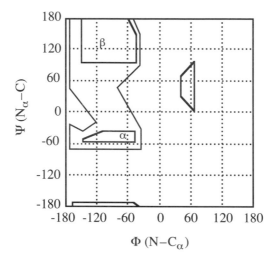

Figure 8.2 Ramachandran diagram for nonglycine amino-acid residues in proteins. Angles Φ and Ψ are as defined in Fig. 8.1.

Ramachandran diagram and plotting each residue number at the position (Φ,Ψ). Refinement papers often include the diagram, with an explanation of any residues that lie in high energy ("forbidden") areas. For an example, see Fig. 8.6 in Section II.C. Glycines, because they lack a side chain, usually account for most of the residues that lie outside allowed regions. If nonglycine residues exhibit forbidden conformational angles, there should be some explanation in terms of structural constraints that overcome the energetic cost of an unusual backbone conformation.

The conformations of amino-acid side chains are unrestrained during refinement. In well-refined models, side-chain single bonds end up in staggered conformations.

B. Resolution and precision of atomic positions

In microscopy, the phrase "resolution of 2 Å" implies that we can resolve objects that are 2 Å apart. If this phrase had the same meaning for a crystallographic model of a protein, in which bond distances average about 1.5 Å, we would be unable to distinguish or resolve adjacent atoms in a 2-Å map. Actually, for a protein refined at 2-Å resolution to an R-factor near 0.2, the situation is much better than the resolution statement seems to imply.

In x-ray crystallography, "2-Å model" means that analysis included reflections out to a distance in the reciprocal lattice of $1/(2\ \text{Å})$ from the center of the diffraction pattern. This means that the model takes into account diffraction from sets of equivalent, parallel planes spaced as closely as 2 Å in the unit cell. (Presumably, data farther out than the stated resolution were unobtainable or too weak to be reliable.) Although the final 2-Å map, viewed as an empty contour surface, may indeed not allow us to discern adjacent atoms, the structural constraints on the model, the requirement that it have reasonable bond lengths and angles, stereochemistry, and conformations, greatly increase the precision of atom positions.

Crystallographers use the Luzzati plot (Fig. 8.3) to estimate the precision of atom locations in a refined crystallographic model. The numbers to the right of each smooth curve on the Luzzati plot are theoretical estimates of the average uncertainty in the positions of atoms in the refined model (more precisely, the rms errors in atom positions). The average uncertainty has been shown to depend on R-factors derived from the final model in various resolution ranges. To use the Luzzati plot, we separate the intensity data into groups of reflections in narrow ranges of $1/d$ (where d is the spacing of real-lattice planes). Then we plot each R-factor (vertical axis) versus the midpoint value of $1/d$ for that group of reflections (horizontal axis). For example, we calculate R using only reflections corresponding to the range $1/d – 0.395\ 0.405$, (reflections in the 2.53–2.47 Å range), and plot this R-factor versus $1/d = 0.400\ \text{Å}^{-1}$, the midpoint value for this group. We repeat this process for the range $1/d = 0.385–0.395$, and so forth. As the theoretical curves indicate, the R-factor typically increases for lower-resolution data (higher values of $1/d$). The resulting curve should roughly fit one

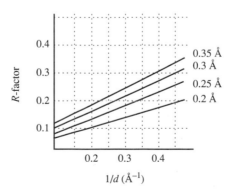

Figure 8.3 Luzzati diagram.

of the theoretical curves on the Luzzati plot. From the theoretical curve closest to the experimental R-factor curve, we learn the average uncertainty in the atom positions of the final model.

Publications of refined structures often include a Luzzati plot, allowing the reader to assess the average uncertainty of atom positions in the model. For highly refined models, rms errors of 0.15 Å are sometimes attained. In Fig. 8.5a, Section II.B, the jagged curve represents the data for the refined model of the protein ALBP. The position of the curve on the Luzzati plot indicates that rms error for this model is about 0.34 Å, about one-fifth the length of a carbon–carbon bond.

In crystallography, unlike microscopy, the term *resolution* simply refers to the amount of data ultimately phased and used in the structure determination. The precision of atom positions depends in part on the resolution limits of the data, but also depends critically on the quality of the data, as reflected by the R-factor. Good data can yield atom positions that are precise to within one-fifth to one-tenth of the stated resolution.

C. Vibration and disorder

Notice, however, that the preceding analysis gives an average, or rms value, of position errors, and further, that the errors result from the limits of accuracy in the data. There are also two important physical (as opposed to statistical) reasons for uncertainty in atom positions: thermal motion and disorder. *Thermal motion* refers to vibration of an atom about its rest position. *Disorder* refers to atoms or groups of atoms that do not occupy the same position in every unit cell, in every asymmetric unit, or in every molecule within an asymmetric unit. In theory, the temperature factor B_j obtained during refinement reflects thermal motion of atom j, while the occupancy n_j reflects disorder. In practice, it is difficult to sort out these two sources of uncertainty.

Occupancies are usually constrained at 1.0 early in refinement, and in many refinements are never released, so that both thermal motion and disorder show their effects on the final B values. In some cases, after refinement converges, a few B values fall far outside the average range for the model. This is sometimes an indication of disorder. Careful examination of $2\mathbf{F}_o - \mathbf{F}_c$ and $\mathbf{F}_o - \mathbf{F}_c$ maps may give evidence for more than one conformation in such a troublesome region. If so, inclusion of multiple conformations followed by refinement of their occupancies may improve the R-factor and the map, revealing the nature of the disorder more clearly.

Assuming that occupancies are correct, B_j is a measure of thermal motion at atom j. In the simplest case of purely harmonic thermal motion of equal magnitude in all directions (called *isotropic* vibration), B_j is related to the magnitude of vibration as follows:

$$B_j = 8\pi^2 \{u_j^2\} = 79 \{u_j^2\} \tag{8.1}$$

where $\{u_j^2\}$ is the mean-square displacement of the atom from its rest position. Thus if the measured B_j is 79 Å^2, the total mean-square displacement of atom j due to vibration is 1.0 Å^2, and the rms displacement is the square root of $\{u_j^2\}$, or 1.0 Å. The B values of 20 Å^2 and 5 Å^2 correspond to rms displacements of 0.5 Å and 0.25 Å.

With small molecules, it is usually possible to obtain anisotropic temperature factors during refinement, giving a picture of the preferred directions of vibration for each atom. But a description of anisotropic vibration requires six parameters per atom, vastly increasing the computational task. In many cases, the total number of parameters sought, including three atomic coordinates, one occupancy, and six thermal parameters per atom, approaches or exceeds the number of measured reflections. As mentioned earlier, for refinement to succeed, observations (measured reflections and constraints such as bond lengths) must outnumber the desired parameters, so that least-squares solutions are adequately overdetermined. For this reason, anisotropic temperature factors for proteins are seldom obtained.

Publications of refined structures often include a plot of average isotropic B values for side-chain and main-chain atoms of each residue, like that shown in Section II.B, Fig. 8.5b for ALBP. Less commonly, pictures of the model are color-coded by temperature factor: red ("hot") for high and blue ("cold") for low values of B. Either presentation calls the user's attention to parts of the molecule that are vibrationally active and parts that are particularly rigid. Not surprisingly, side-chain temperature factors are larger and more varied (5–60 Å^2) than those of main-chain atoms (5–35 Å^2). Values of B greater than 60 Å^2 may signify disorder or errors in the model.

Remember that we see in a crystallographic model an average of all the molecules that diffracted the x-rays. Furthermore, we see a static structure representing a stable conformation of a dynamic molecule. It is sobering to realize that the crystallographic model of ALBP exhibits no obvious path for entry and departure of its ligands, lipid molecules such as oleic acid. Similarly, comparison of the crystallographic models of hemoglobin and deoxyhemoglobin reveals no path for entry of the tiny O_2 molecule. Seemingly simple processes like the binding of small ligands to proteins often involve conformational changes to states not revealed by crystallographic analysis.

Nevertheless, the crystallographic model contributes importantly to solving such problems of molecular dynamics. The refined structure serves as a starting point for simulations of molecular motion. From that starting point, which undoubtedly represents one common conformation of the protein, and from the equations of motion of atoms in the force fields of electrostatic and van der Waals forces, scientists can calculate the normal vibrational motions of the molecules and can simulate random molecular motion, thus gaining insights into how conformational change gives rise to biomolecular function. Even though the crystallographic model is static, it is an essential starting point in revealing the dynamic aspects of structure.

D. Other limitations of crystallographic models

The limitations discussed so far apply to all models, and suggest questions that the user of crystallographic results should ask routinely. Other limitations are special cases that may or may not apply to a given model. It is important to read the original publications of a structure to see whether any of the following limitations apply.

Low-resolution models

Not all published models are refined to high resolution. For instance, publication of a low-resolution structure may be warranted if it displays an interesting and suggestive arrangement of cofactors or clusters of metal ions, provides possible insights into conformations of a new family of proteins, or displays the application of new imaging methods. In some cases, the published structure is only a crude electron-density model. Or perhaps it contains only the estimated positions of α-carbons. Such models may be of limited use for comparison with other proteins, but of course, they cannot support detailed molecular analysis. In α-carbon models, there is great deal of uncertainty in the positions, and even in the number, of α-carbons. Often, further refinement of these models reveals errors in the chain tracing. Protein Data Bank header information includes the model resolution and descriptions of its contents. Coordinate entries in the PDB list are marked *a* if they contain only α-carbon coordinates.

Disordered regions

Occasionally, portions of a protein are never found in the electron-density maps, presumably because the region is highly disordered or in motion and thus invisible on the time scale of crystallography. It is not at all uncommon for residues at termini, especially the N-terminus, to be missing from a model. Discussions of these structure-specific problems are included in a thorough refinement paper, as well as in PDB header information.

Unexplained density

Just as the auto mechanic sometimes has parts left over, electron-density maps occasionally show clear, empty density after all known contents of the crystal have been located. Among possible explanations are ions like phosphate and sulfate from the mother liquor; reagents like mercaptoethanol, dithiothreitol, or detergents used in purification or crystallization; or cofactors, inhibitors, allosteric effectors, or other small molecules that survived the protein purification. Later discovery of previously unknown but important ligands has sometimes resulted in subsequent interpretation of empty density.

Distortions due to crystal packing

Refinement papers should also mention any evidence that the protein is affected by crystallization. Packing effects may be evident in the model itself. For example, packing may induce slight differences between symmetrically related subunits within an asymmetric unit. Examination of the neighborhood around such differences may reveal that intermolecular contact is a possible cause. In areas where subunits come into direct contact or close contact through intervening water, surface temperature factors are usually lower than at other surface regions.

E. Summary

Sensible use of a crystallographic model, like any complex tool, requires understanding of its limitations. Some limitations, like the precision of atom positions and the static nature of the model, are general constraints on use. Others, like disordered regions, undetected portions of sequence, unexplained density, and packing effects, are model-specific. If you use a protein model from the PDB without reading the header information, or without reading the original publications, you may be missing something vital to appropriate use of the model. The result may be no more than a crash of your graphics software because of unexpected input like a file containing only α-carbons. Or more seriously, you may devise and publish a detailed molecular explanation based on a structural feature that is quite uncertain. In some cases, the model is not enough. If specific structural details of the model are crucial to a proposed mechanism or explanation, it is advisable to look at the electron density map in the important region, and be sure that the map is well defined there, and that the model fits it well.

III. Reading a crystallography paper

A. Introduction

Judging the quality and potential usefulness of a crystallographic model means first extracting the criteria of quality from published reports. To help you begin to develop this skill, I will walk you through an attempt to cull such information from publications of a "typical" crystallographic project. Following are annotated portions of two recent papers reporting the crystallization and structure determination of adipocyte lipid binding protein (ALBP), a member of a family of hydrophobic ligand-binding proteins. The first paper[1] appeared in August 1991, announcing the purification and crystallization of the protein, and presenting preliminary results of crystallographic analysis. The second paper[2], published in April 1992, presented the completed structure with experimental details. In the following sections, I will focus primarily on the experimental and results sections of the papers, and specifically on (1) methods and concepts treated earlier in this book and (2) criteria of refinement convergence and quality of the model.

[1] Z. Xu, M. K. Buelt, L. J. Banaszak, and G. A. Bernlohr, Expression, purification, and crystallization of the adipocyte lipid binding protein, *Journal of Biological Chemistry* **266**, 14367–14370, 1991.
[2] Z. Xu, D. A. Bernlohr, and L. J. Banaszak, Crystal structure of recombinant murine adipocyte lipid-binding protein, *Biochemistry* **31**, 3484–3492, 1992.

Although I have reproduced parts of the published experimental procedures here (with the permission of the authors and publishers), you may wish to obtain these papers from your library and read them before proceeding with this example. See Footnotes 1 and 2 (on the previous page) for complete references.

In the following material, sections taken from the original papers are presented in smaller type. Annotations are in the usual type size. For convenience, figures and tables are renumbered in sequence with those of this chapter. For access to references cited in excerpts, see the complete papers. Stereo illustrations of maps and models (not part of the papers) are derived from files kindly provided by Zhaohui Xu. I am indebted to Xu and to Leonard J. Banaszak for allowing me to use their work as an example, and for supplying me with a complete reconstruction of this recent structure determination project.

B. Annotated excerpts of the preliminary (8/91) paper

All reprinted parts of this paper (cited in Footnote 1) appear with the permission of Professor Leonard J. Banaszak and the American Society for Biochemistry and Molecular Biology, Inc., publisher of the *Journal of Biological Chemistry*.

In the August 5, 1991 issue of *Journal of Biological Chemistry*, Xu, Buelt, Banaszak, and Bernlohr reported the cloning, expression, purification, and crystallization of adipocyte lipid binding protein (ALBP, or rALBP for the recombinant form), along with preliminary results of crystallographic analysis. This type of paper typically appears as soon as a research team has carried the structure project far enough to know that it promises to produce a good model. An important aim of announcing that work is in progress on a molecule is to avoid duplication of effort in other laboratories. While one might cynically judge that such papers constitute a defense of territory, and a grab for priority in the work at hand, something much more important is at stake. Crystallographic structure determination is a massive and expensive undertaking. The worldwide resources, both equipment and qualified scientists, for structure determination are inadequate for the many molecules we would like to understand. Duplication of effort on the same molecule squanders limited resources in this important field. So generally, as soon as a team has good evidence that they can produce a structure, they alert the crystallographic community to prevent parallel work from beginning in other labs.

The following paragraph is excerpted from the preliminary (8/91) paper, "Experimental Procedures" section:

> *Crystallization*–Small crystals (0.05 × 0.1 ×01 mm) were obtained using the hang-
> ing drop/vapor equilibrium method (18). 10-µl drops of 2.5 mg/ml ALBP in
> 0.05 M Tris, 60% ammonium sulfate, 1 mM EDTA, 1 mM dithiothreitol, 0.05% so-
> dium azide buffer with a pH of 7.0 (crystallization buffer) were suspended over
> wells containing the same buffer with varying concentrations of ammonium sulfate,
> from 75 to 85% saturation. Small, well shaped crystals were formed within a month
> at an 80% saturation and 19°C. These crystals were isolated, washed with mother
> liquid, and used as seeds by transferring them into a 10-µl drop of 4 mg/ml fresh
> ALBP in the 80% saturation crystallization buffer over a well containing the same
> buffer. Large crystals, 0.3 × 0.4 × 0.4 mm, grew in 2 days at a constant temperature
> of 19°C.

The precipitant used here is ammonium sulfate, which precipitates proteins by salting out. Notice that Xu and co-workers tried a range of precipitant concentrations, probably after preliminary trials over a wider range.
Crystals produced by the hanging-drop method (Chapter 3, Section III.B)
were too small for x-ray analysis but were judged to be of good quality. The
small crystals were used as seeds on which to grow larger crystals under the
same conditions that produced the best small crystals. This method, called
repeated seeding, was also discussed in Chapter 3. The initial unseeded
crystallization probably fails to produce large crystals because many crystals form at about the same rate, and the supply of soluble protein is depleted before any crystals become large. The seeded crystallization is probably effective because it decreases the number of sites of crystal growth,
causing more protein to go into fewer crystals. Notice also how much faster
crystals grow in the seeded drops (two days) than in the unseeded (one
month). The preformed crystals provide nucleation sites for immediate further growth, whereas the first crystals form by random nucleation events
that are usually rate-limiting in unseeded crystallizations.

> *Data Collection and Processing*–Crystals were analyzed with the area detector dif-
> fractometer from Siemens/Nicolet. A 0.8-mm collimator was used, and the crystal
> to detector distance was set at 12 cm with the detector midpoint at $2\theta = 15°$. One
> ϕ scan totaling 90° and three Ω scans of 68° with χ at 45° were collected with the
> Rigaku Ru200 operating at 50 kV and 180 mA. Each frame consisted of a 0.25° ro-
> tation taken for 120 s. The diffractometer data were analyzed with the Xengen
> package of programs. Raw data within 50 frames were searched to find about 100
> strong reflections which were then indexed, and the cell dimensions were refined
> by least squares methods. Data from different scans were integrated separately and
> then merged together.

The angles ϕ, χ, ω, and 2θ refer to the diffractometer angles described in Chapter 4, Section III.D and shown in Fig. 4.21. The Rigaku Ru200 is the x-ray source, a rotating anode tube. Each frame of data collection is, in essence, one electronic film on which are recorded all reflections that pass through the sphere of reflection during a 0.25° rotation of the crystal. This rotation size is chosen to collect as many reflections as possible without overlap. As mentioned in Chapter 4, diffractometer measurements are almost fully automated. In this instance, cell dimensions were worked out by a computer program that finds 100 strong reflections and indexes them. Then the program employs a least-squares routine (Chapter 7, Section VI.A) to refine the unit-cell dimensions, by finding the cell lengths and angles that minimize the difference between the actual positions of the 100 test reflections and the positions of the same reflections as calculated from the current trial set of cell dimensions. (Least-squares procedures are used in many areas of crystallography in addition to structure refinement.) Using accurate cell dimensions, the program indexed all reflections, and then integrated the x-ray counts received at each location to obtain reflection intensities.

The following excerpt is from the "Results and Discussion" section of the 9/91 paper:

> Crystallization experiments using rALBP were immediately successful. With seeding, octahedral crystals of the *apo*-protein grow to a length of 0.4 mm and a height of 0.3 mm. These crystals give diffraction data to 2.4 Å. An entire data set was collected to 2.7-Å resolution using the area detector system. Statistical details of the combined x-ray data set are presented in Table 8.1.

Xu and colleagues had exceptionally good fortune in obtaining crystals. Efforts to crystallize a desirable protein can give success in a few weeks, or never, or anything in between. The time required here exceeds even very optimistic hopes. The extent of diffraction in preliminary tests (2.4 Å) is a key indicator that the crystals might yield a high-quality structure.

Table 8.1 gives the reader a glimpse into the quality of the native data set. The 0.25° frames of data from the area detector are merged into one data set by multiplying all intensities in each frame by a scale factor. A least-squares procedure determines scale factors that minimize the differences between intensities of identical reflections observed on different frames. The merging R-factor [see Equation (7.10)] gives the level of agreement among the different frames of data after scaling. In this type of R-factor, $|\mathbf{F}_{obs}|$'s are derived from averaged, scaled intensities for all observations of one reflection, and corresponding $|\mathbf{F}_{calc}|$'s are derived from scaled intensi-

Table 8.1

X-ray Data Collection Statistics for Crystalline ALBP

Merging R-factor based on I	0.0426
Resolution limits	2.2 Å
Number of observations	20,478
Number of unique x-ray reflections collected	5,473
Average number of observations for each reflection	4.0
% of possible reflections collected to 2.7 Å	98
% of possible reflections collected to 2.4 Å	36

ties for individual observations of the same reflection. The better the agreement between these two quantities throughout the data set, the lower the R-factor. In this case, individual scaled intensities agree with their scaled averages to within about 4%.

You can see from Table 8.1 that 98% of the reflections available out to 2.7 Å [those lying within a sphere of radius $1/(2.7$ Å$)$ centered at the origin of the reciprocal lattice] were measured, and on the average, each reflection was measured four times. Additional reflections were measured out to 2.4 Å. The number of available reflections increases with the third power of the radius of the sampled region in the reciprocal lattice, so a seemingly small increase in resolution from 2.7 to 2.4 Å requires 40% more data. [Compare $(1/2.4)^3$ with $(1/2.7)^3$]. For a rough calculation of the number of available reflections at specified resolution, see annotations of the 4/92 paper, below.

The lattice type was orthorhombic with unit cell dimension of $a = 34.4$ Å, $b = 54.8$ Å, $c = 76.3$ Å. The x-ray diffraction data were examined for systematic absences to determine the space group. Such absences were observed along the **a***, **b***, and **c*** axes. Only reflections with h, k, or $l = 2n$ were observed along the reciprocal axes. This indicated that the space group is $P2_12_12_1$. A unit cell with the dimensions described above has a volume of 1.44×10^5 Å^3. Assuming that half of the crystal volume is water, the volume of protein is approximately 7.2×10^4 Å^3. Considering the space group here, the volume protein in 1 asymmetric unit would be 1.8×10^4 Å^3. By averaging the specific volume of constituent amino acids, the specific volume of ALBP is 0.715 ml/g. This led to the conclusion that the molecular mass in one asymmetric unit is 15,155 daltons. Since the molecular mass of ALBP is approximately 15 kDa, there is only 1 molecule of ALBP in an asymmetric unit.

Recall from Chapter 5, Section IV.C that for a twofold screw axis along the c edge, all odd-numbered $00l$ reflections are absent. In the space group $P2_12_12_1$, the unit cell possesses twofold screw axes on all three edges, so odd-numbered reflections on all three principal axes of the reciprocal lattice

($h00$, $0k0$, and $00l$) are missing. The presence of only even-numbered reflections on the reciprocal-lattice axes announces that the ALBP unit cell has $P2_12_12_1$ symmetry.

As described in Chapter 3, Section IV, the number of molecules per asymmetric unit can be determined from unit-cell dimensions and a rough estimate of the protein/water ratio. Since this number is an integer, even a rough calculation can give a reliable answer. The assumption that ALBP crystals are 50% water is no more than a guess taken from near the middle of the range for protein crystals (30–78%). The unit-cell volume is (34.4 Å)(54.8Å)(76.3Å) = 1.44×10^5 Å^3, and if half that volume is protein, the protein volume is 7.2×10^4 Å^3. In space group $P2_12_12_1$, there are four equivalent positions (Chapter 4, Section II.H), so there are four asymmetric units per unit cell. Each one must occupy one-fourth of the protein volume, so the volume of the asymmetric unit is one-fourth of 7.2×10^4, or 1.8×10^4 Å^3. The stated specific volume (volume per gram) of the protein is the weighted average of the specific volumes of the amino-acid residues (which can be looked up), weighted according to the amino-acid composition of ALBP. The molecular mass of one asymmetric unit is obtained by converting the density of ALBP in grams per milliliter (which is roughly the inverse of the specific volume) to daltons per cubic angstrom, and then multiplying by the volume of the asymmetric unit, as follows:

$$\frac{1 \text{ g}}{0.715 \text{ ml}} \cdot \frac{1 \text{ ml}}{\text{cm}^3} \cdot \frac{\text{cm}^3}{(10^8)^3 \text{Å}^3} \cdot \frac{6.02 \times 10^{23} \text{daltons}}{\text{g}} \cdot (1.8 \times 10^4) \text{ Å}^3$$

$$= 1.5 \times 10^4 \text{ daltons} \tag{8.2}$$

This result is very close to the known molecular mass of ALBP, so there is one ALBP molecule per asymmetric unit. This knowledge is an aid to early map interpretation.

The excerpt from "Results and Discussion" continued:

> As indicated, ALBP belongs to a family of low molecular weight fatty acid binding proteins. The sequences of the proteins in the family have been shown to be very similar and in particular in the amino-terminal domain where Y19[3] resides. Among them, the structure of myelin P2 and IFABP has been solved. Since the amino acid identity between ALBP and myelin P2 is about 69%, P2 should be a good starting structure to obtain phase information for ALBP using the method of molecular replacement. Preliminary solutions to the rotation and translation functions have been obtained. Seeding techniques will allow us to obtain large crystals for further study of the *holo*– and phosphorylated protein. By comparing the crystal structures of

---

[3] Y19 is tyrosine 19, a residue considered important to the function of ALBP.

these different forms, it should be possible to structurally determine the effects of protein phosphorylation on ligand binding and ligand binding on phosphorylation.

Because ALBP is related to several proteins of known structure, molecular replacement is an attractive option for phasing. The choice of a phasing model is simple here: just pick the one with amino-acid sequence most similar to ALBP, which is myelin P2 protein. Solution of rotation and translation functions refers to the search for orientation and position of the phasing model (P2) in the unit cell of ALBP. The subsequent paper provides more details.

C. Annotated excerpts from the full structure-determination (4/92) paper

All reprinted parts of this paper (cited in Footnote 2, above) appear with the permission of Professor Leonard J. Banaszak and the American Chemical Society, publisher of *Biochemistry*.

In April 1992, the structure-determination paper appeared in *Biochemistry*. This paper contains full description of the experimental work, and a complete analysis of the structure. The following is from the 4/92 paper, "Abstract" section:

> Adipocyte lipid-binding protein (ALBP) is the adipocyte member of an intracellular hydrophobic ligand-binding protein family. ALBP is phosphorylated by the insulin receptor kinase upon insulin stimulation. The crystal structure of recombinant murine ALBP has been determined and refined to 2.5 Å. The final R-factor for the model is 0.18 with good canonical properties.

A 2.5-Å model refined to an R-factor of 0.18 should be a detailed model. "Good canonical properties" means good agreement with accepted values of bond lengths, bond angles, and planarity of peptide bonds.

The following is an excerpt from the "Materials and Methods" section of the 4/92 paper:

> *Crystals and X-ray Data Collection.* Detailed information concerning protein purification, crystallization, and X-ray data collection can be found in a previous report (Xu *et al.*, 1991) and will be mentioned here in summary form. Recombinant murine *apo*-ALBP crystallizes in the orthorhombic space group $P2_12_12_1$ with the following unit cell dimensions: $a = 34.4$ Å, $b = 54.8$ Å, and $c = 76.3$ Å. The asymmetric unit contains one molecule with a molecular weight of 14,500. The entire diffraction data set was collected on one crystal. In the resolution range $\infty - 2.5$

Å, 5115 of the 5227 theoretically possible reflections were measured. Unless otherwise noted the diffraction data with intensities greater than 2σ were used for structure determination and refinement. As can be seen in Table 8.2, this included about 96% of the measured data.

This section reviews briefly the results of the preliminary paper. In the early stages of the work, reflections weaker than twice the standard deviation for all reflections (2σ) were omitted from Fourier syntheses, because of greater uncertainty in the measurements of weak reflections. Table 8.2 is discussed below.

The diffractometer software computes the number of reflections available at 2.5-Å resolution by counting the number of reciprocal-lattice points that lie within a sphere of radius ($1/[2.5$ Å]), centered at the origin of the reciprocal lattice. This number is roughly equal to the number of reciprocal unit cells within the $1/[2.5$ Å] sphere, which is, again roughly, the volume of the sphere (V_{rs}) divided by the volume of the reciprocal unit cell (V_{rc}). The volume of the reciprocal unit cell is the inverse of the real unit-cell volume V. So the number of reflections available at 2.5-Å resolution is approximately (V_{rs}) · (V). Because of the symmetry of the reciprocal lattice and of the $P2_12_12_1$ space group, only one-eighth of the reflections are unique (Chapter 4, Section III.G). So the number of unique reflections is approximately (V_{1s}) · (V)/8, or

$$\frac{\frac{4}{3}\pi\left(\dfrac{1}{2.5\,\text{Å}}\right)^3 (1.44 \times 10^5\,\text{Å}^3)}{8} = 4825 \text{ reflections}$$

The 8% difference between this result and the stated 5227 reflections is due to the approximations made here, and to the sensitivity of the calculation to small round-off in unit-cell dimensions.

Molecular Replacement. The tertiary structure of crystalline ALBP was solved by using the molecular replacement method incorporated into the XPLOR computer program (Brunger *et al.*, 1987). The refined crystal structure of myelin P2 protein without solvent and fatty acid was used as the probe structure throughout the molecular replacement studies. We are indebted to Dr. A. Jones and his colleagues for permission to use their refined P2 coordinates before publication.

Note that the myelin P2 coordinates were not yet available from the Protein Data Bank and were obtained directly from the laboratory in which the P2 structure was determined. Because of the time required for publication of research papers and processing of coordinates by the PDB, coordinates may be available directly from a crystallographic research group one or two years before they are available from PDB.

In this project, the search for the best orientation and position of P2 in the ALBP unit cell was divided into three parts: a rotation search to find promising orientations, refinement of the most promising orientations to find the best orientation, and a translation search to find the best position. Here are the details of the search:

> (1) *Rotation Search.* The rotation search was carried out using the Patterson search procedures in XPLOR. The probe Patterson maps were computed from structure factors calculated by placing the P2 coordinates into an orthorhombic cell with 100-Å edges. One thousand highest Patterson vectors in the range of 5–15 Å were selected and rotated using the pseudoorthogonal Eulerian angles (θ_+, θ_2, θ_-) as defined by Lattman (1985). The angular search interval for θ_2 was set to 2.5°; intervals for θ_+ and θ_- are functions of θ_2. The rotation search was restricted to the asymmetric unit $\theta_- = 0–180°$, $\theta_2 = 0–90°$, $\theta_+ = 0–720°$ for the $P2_12_12_1$ space group (Rao, *et al.*, 1980). XPLOR produces a sorted list of the correlation results simplifying final interpretation (Brunger 1990).

XPLOR is a modern package of refinement programs that includes powerful procedures for energy refinement by simulated annealing, in addition to more traditional tools like least-squares methods and molecular replacement searches. The package is available for use on many different computer systems. Simulated annealing for large molecules usually requires supercomputers.

The P2 phasing model is referred to here as the *probe*. For the rotation search, the probe was placed in a unit cell of arbitrary size and $\mathbf{F}_{calc}$'s were obtained from this molecular model, using Equation (5.15). Then a Patterson map was computed from these $\mathbf{F}_{calc}$'s using Equation (6.10). Recall that Patterson maps reflect the molecule's orientation, but not its position. All peaks in the Patterson map except the strongest 1000 were eliminated. Then the resulting simplified map was compared to a Patterson map calculated from ALBP reflection intensities. The probe Patterson was rotated in a three-dimensional coordinate system to find the orientation that best fit the ALBP Patterson. (The angles refer to a standard set of angles for rotating the model through all unique orientations.) A plot of the angles versus some criterion of coincidence between peaks in the two Patterson maps is called a *rotation function*. Peaks in the rotation function occur at sets of angles where many coincidences occur. The coincidences are not perfect, because there is a finite interval between angles tested, and the best orientation is likely to lie between test angles. The interval is made small enough to avoid missing promising orientations altogether.

(2) *Patterson Correlation Refinement.* To select which of the orientations determined from the rotation search is the correct solution a Patterson correlation refinement of the peak list of the rotation function was performed. This was carried out by minimization against a target function defined by Brunger (1990) and as implemented in XPLOR. The search model P2 was optimized for each of the selected peaks of the rotation function.

As discussed later in the "Results" section, the rotation function contains many peaks. The strongest 100 peaks are selected and each orientation is refined by least squares to produce the best fit to the ALBP Patterson map. For each refined orientation, a correlation coefficient is computed. The orientation giving the highest correlation coefficient is chosen as the best orientation for the phasing model.

(3) *Translation Search.* A translation search was done by using the P2 probe molecule oriented by the rotation function studies and refined by the Patterson correlation method. The translation search employed the standard linear correlation coefficient between the normalized observed structure factors and the normalized calculated structure factors (Funinaga & Read, 1987; Brunger, 1990). X-ray diffraction data from 10–3-Å resolution were used. Search was made in the range $x = 0–0.5$, $y = 0–0.5$, and $z = 0–0.5$, with the sampling interval 0.0125 of the unit cell length.

The last step in molecular replacement is to find the best position for the probe molecule in the ALBP unit cell. The P2 orientation obtained from the rotation search and refinement is tried in all unique locations at intervals of one-eightieth of the unit-cell axis lengths. The symmetry of the $P2_12_12_1$ unit cell allows this search to be confined to the region bound by half of each cell axis. The total number of positions tested is thus (40)(40)(40) or 64,000. For each position, F_{calc}'s are computed [Equation (5.15)] from the P2 model and their amplitudes are compared with the $|F_{obs}|$'s from the ALBP native data set. An unspecified correlation coefficient, probably similar to an R-factor, is computed for each P2 position, and the position giving P2 $|F_{calc}|$'s in best agreement with ALBP $|F_{obs}|$'s is chosen as the best position for P2 as a phasing model. The starting phase estimates for the refinement were thus the phases of F_{calc}'s computed [Equation (5.15)] from P2 in the final orientation and position determined by the three-stage molecular replacement search.

Structure Refinement. The refinement of the structure was based on an energy function approach (Brunger *et al.*, 1987): arbitrary combinations of empirical and effective energy terms describing crystallographic data as implemented in XPLOR. Molecular model building was done on an IRIS Workstation (Silicon Graphics) with the software TOM, a version of FRODO (Jones, 1978).

The initial model of ALBP was built by simply putting the amino acid sequence of ALBP into the molecular structure of myelin P2 protein. After a 20-step rigid-body refinement of the positions and orientations of the molecule, crystallographic refinement with simulated annealing was carried out using a slow-cooling protocol (Brunger *et al.*, 1989, 1990). Temperature factor refinement of grouped atoms, one for backbone and one for side-chain atoms for each residue, was initiated after the *R*-factor dropped to 0.249.

The first electron-density map was computed [Equation (7.3)] with $|\mathbf{F}_{obs}|$'s from the ALBP data set and α_{calc}'s from the oriented P2 molecule. Plate 10 shows a small section of this map superimposed on the final model.

An early map like Plate 10, computed from initial phase estimates, harbors many errors, where the map does not agree with the model ultimately derived from refinement. In this section, you can see both false breaks and false connections in the density. For example, there are breaks in density at C_β of the phenylalanine residue (side chain ending with six-membered ring) on the right, and along the protein backbone at the upper left. The lobe of density corresponding to the valine side chain (center front) is disconnected and out of place. There is a false connection between density of the carbonyl oxygen (red) at lower left and side-chain density above. Subsequent refinement is aimed at improving this map.

Next, the side chains of P2 were replaced with the side chains of ALBP at corresponding positions in the amino-acid sequence to produce the first ALBP model. The position and orientation of this model were refined by least squares, treating the model as a rigid body. Subsequent refinement was by simulated annealing. At first, all temperature factors were constrained at 15.0 Å^2. After the first round of simulated annealing, temperature factors were allowed to refine for atoms in groups, one value of *B* for all backbone atoms, and another for side-chain atoms.

The new coordinates were checked and adjusted against a $(2|\mathbf{F}_o| - |\mathbf{F}_c|)$ and a $(|\mathbf{F}_o| - |\mathbf{F}_c|)$ electron density map, where $|\mathbf{F}_o|$ and $|\mathbf{F}_c|$ are the observed and calculated structure factor amplitudes. Phases are calculated from the crystal coordinates. The Fourier maps were calculated on a grid corresponding to one-third of the high-resolution limit of the input diffraction data. All residues were inspected on the graphics system at several stages of refinement. The adjustments were made on the basis of the following criteria: (a) that an atom was located in low electron density in the $(2|\mathbf{F}_o| - |\mathbf{F}_c|)$ map or negative electron density in the $(|\mathbf{F}_o| - |\mathbf{F}_c|)$ map; (b) that the parameters for the Φ, Ψ angles placed the residue outside the acceptable regions in the Ramachandran diagram. Iterative refinement and model adjustment against a new electron density map was carried out until the *R*-factor appeared unaffected. Isotropic temperature factors for individual atoms were then included in the refinement.

In between rounds of computerized refinement, maps were computed using $|\mathbf{F}_{obs}|$'s from the ALBP data set and α_{calc}'s from the current model [taken from $|\mathbf{F}_{calc}|$'s computed by Equation (5.15)]. The model was corrected where the fit to maps was poor, or where the Ramachandran angles Φ and Ψ were forbidden. Notice that the use of $2\mathbf{F}_o - \mathbf{F}_c$ and $\mathbf{F}_o - \mathbf{F}_c$ maps [Equations (7.4) and (7.5)] is as described in Chapter 7, Section IV.B. When alternating rounds of refinement and map fitting produced no further improvement in R-factor, temperature factors for each atom were allowed to refine individually, leading to further decrease in R.

> The next stage of the crystallographic study included the location of solvent molecules. They were identified as well-defined peaks in the electron-density maps within hydrogen-bonding distance of appropriate protein atoms or another solvent atoms. Solvent atoms were assigned as water molecules and refined as oxygen atoms. Those that refined to positions too close to other atoms, ended up located in low electron density, or had associated temperature factors greater than 50 Å^2 were removed from the coordinate list in the subsequent stage. The occupancy for all atoms, including solvent molecules, was kept at 1.0 throughout the refinement. Detailed progress of the crystallographic refinement is given in Table 8.2.

Finally, ordered water molecules were added to the model where unexplained electron density was present in chemically feasible locations for water molecules. Temperature factors for these molecules (treated as oxygen atoms) were allowed to refine individually. If refinement moved these molecules into unrealistic positions or increased their temperature factors excessively, the molecules were deleted from the model. Occupancies were constrained to 1.0 throughout the refinement. This means that B values reflect both thermal motion and disorder (Section II.C). Because all B values fall into a reasonable range, the variation in B can be attributed to thermal motion. Table 8.2 shows the progress of the refinement.

Note that R drops precipitously in the first stages of refinement after ALBP side chains replace those of P2. Note also that R and the deviations from ideal bond lengths, bond angles, and planarity of peptide bonds decline smoothly throughout the later stages of refinement. The small increase in R at the end is due to inclusion of weaker reflections in the final round of simulated annealing.

The following excerpt is from the "Results" section of the 8/92 paper:

> *Molecular Replacement.* From the initial rotation search, the 101 highest peaks were chosen for further study. These are shown in Fig. 8.4. The highest peak of the rotation function had a value 4.8 times the standard deviation above the mean and 1.8 times the standard deviation above the next highest peak. The orientation was consistently the highest peak for diffraction data within the resolution ranges 10–5, 7–5, and 7–3 Å. Apart from peak number 1, six strong peaks emerged after PC[4]

---

[4] Patterson correlation

refinement, as can be seen in Fig. 8.4*b*. These peaks all corresponded to approximately the same orientation as peak number 1. Three of them were initially away from that orientation and converged to it during the PC refinement.

A translation search as implemented in XPLOR was used to find the molecular position of the now oriented P2 probe in the ALBP unit cell. Only a single position emerged at $x = 0.250$, $y = 0.425$, $z = 0.138$ with a correlation coefficient of 0.419. The initial R-factor for the P2 coordinates in the determined molecular orientation and position was 0.470 including X-ray data in the resolution range of 10–3 Å. A rigid-body refinement of orientation and position reduced the starting R-factor to only 0.456, probably attesting to the efficacy of the Patterson refinement in XPLOR.

Table 8.2

Progress of Refinement

Stage *	Number of Reflections	R-factor	B (Å^2)	RMS Deviations			
				Solvent Included	Bond Length (Å)	Bond Angle (deg)	Planarity (deg)
1	2976	0.458	15.0		0.065	4.12	9.015
2	2976	0.456	15.0		0.065	4.12	9.012
3	4579	0.235	group		0.019	3.17	1.506
4							
5	4579	0.220	indiv.		0.018	3.77	1.408
6							
7	4579	0.197	indiv.	31	0.018	3.73	1.366
8							
9	4579	0.172	indiv.	88	0.016	3.47	1.139
10							
11	4773	0.183	indiv.	69	0.017	3.46	1.070

* Key to stages of refinement:

Stage	Action
1	Starting model
2	Rigid-body refinement
3	Simulated annealing
4	Model rebuilt using $(2F_o - F_c)$ and $(F_o - F_c)$ electron density maps
5	Simulated annealing
6	Model rebuilt using $(2F_o - F_c)$ and $(F_o - F_c)$ electron density maps, H_2O included
7	Simulated annealing
8	Model rebuilt using $(2F_o - F_c)$ and $(F_o - F_c)$ electron density maps, H_2O included
9	Simulated annealing
10	Model rebuilt using $(2F_o - F_c)$ and $(F_o - F_c)$ electron density maps, H_2O included
11	Simulated annealing

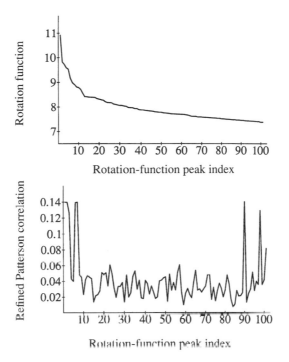

Figure 8.4 Rotation function results: P2 into crystalline ALBP. (*a*) Plot of the 101 best solutions to the rotation function, each peak numbered in the horizontal direction (abscissa). The correlation between the Patterson's of the probe molecule and the measured ALBP x-ray results are shown in the vertical direction (ordinate) and are given in arbitrary units. (*b*) Description of the rotation studies after Patterson correlation refinement. The peak numbers plotted in both panels (*a*) and (*b*) are the same.

In Fig. 8.4*a* , the value of the rotation function, which indicates how well the probe and ALBP Patterson maps agree with each other, is plotted vertically against numbers assigned to the 101 orientations that produced best agreement. Then each of the 101 orientations were individually refined further, by finding the nearby orientation having maximum value of the rotation function. In some cases, different peaks refined to the same final orientation. Each refined orientation of the probe received a correlation coefficient that shows how well it fits the Patterson map of ALBP. The orientation receiving the highest correlation coefficient was taken as the best orientation of the probe, and then used to refine the position of the probe in the ALBP unit cell. The orientation and position of the model obtained from the molecular replacement search were so good that refinement of the model as a rigid body produced only slight improvement in *R*. The authors attribute this to the Patterson correlation refinement of model orientation, stage two of the search.

> *Refined Structure of apo-ALBP.* The refined ALBP structure has a R-factor 0.183 when all observed X-ray data (4773 reflections) between 8.0 and 2.5 Å are included. The rms deviation of bond lengths, bond angle, and planarity from ideality is 0.017 Å, 3.46°, and 1.07°, respectively. An estimate of the upper limit of error in atomic coordinates is obtained by the method of Luzzati (1952). Figure 8.5 summarizes the overall refined model.
>
> The plot presented in Fig. 8.5*a* suggests that the upper limit for the mean error of the refined ALBP coordinates is around 0.34 Å. The mean temperature factors for main-chain and side-chain atoms are plotted in Fig. 8.5*b*.

The final R-factor and structural parameters exceed the standards described in Section I, and attest to the high quality of this model. Atom locations are precise to an average of 0.34 Å, about one-fifth of a carbon–carbon covalent bond length. The plot of temperature factors shows greater variability and range for side-chain atoms, as expected, and shows no outlying values. The model defines the positions of all amino-acid residues in the protein.

> Careful examination of $(2|\mathbf{F}_o| - |\mathbf{F}_c|)$ and $(|\mathbf{F}_o| - |\mathbf{F}_c|)$ maps at each refinement step led to the conclusion that no bound ligand was present. There was no continuous positive electron density present near the ligand-binding site as identified in both P2 (Jones, *et al.*, 1988) and IFABP (Sacchettini *et al.*, 1989a). The absence of bound fatty acid in crystalline ALBP is consistent with the chemical modification experiment which indicates ALBP purified from *E. coli* is devoid of fatty acid (Xu *et al.*, 1991). The final refined coordinate list includes 1017 protein atoms and 69 water molecules.

The final maps exhibit no unexplained electron density. This is of special concern because ALBP is a ligand-binding protein (its ligand is a fatty acid), and ligands sometimes survive purification and crystallization, and are found in the final electron-density map. It is implied by references to "*apo*-protein" and "*holo*-protein" that attempts to determine the structure of an ALBP/ligand complex are under way. If it is desired to detect conformational changes on ligand binding, then it is crucial to know that no ligand is bound to this *apo*-protein, so that conformational differences between *apo*- and *holo*-forms, if found, can reliably be attributed to ligand binding.

To compare *apo*- and *holo*-forms of proteins after both structures have been determined independently, crystallographers often compute difference Fourier syntheses (Chapter 7, Section IV.B), in which each Fourier term contains the structure-factor difference $\mathbf{F}_{holo} - \mathbf{F}_{apo}$. A contour map

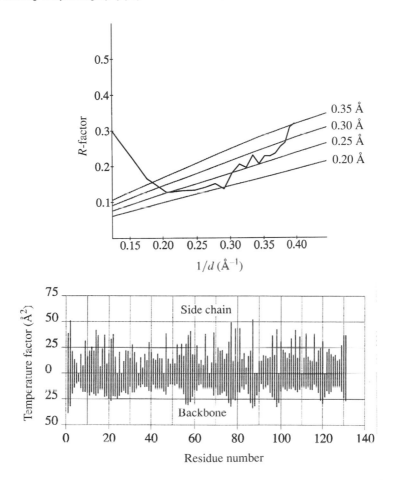

Figure 8.5 ALBP refinement results. (*a*) Theoretical estimates of the rms positional errors in atomic coordinates according to Luzzati (1952) are shown superimposed on the curve for the ALBP diffraction data. The coordinate error estimated from this plot is 0.25 Å with an upper limit of about 0.35 Å. (*b*) Mean values of the main-chain and side-chain temperature factors are plotted against the residue number. The temperature factors are those obtained from the final refinement cycles.

of this Fourier series is called a difference map, and it shows only the differences between the *holo*- and *apo*-forms. Like the $\mathbf{F}_o - \mathbf{F}_c$ map, the $\mathbf{F}_{holo} - \mathbf{F}_{apo}$ map contains both positive and negative "density." Positive density occurs where the electron density of the *holo*-form is greater than that of the *apo*-form, so the ligand shows up clearly in positive density. In addition, conformational differences between *holo*- and *apo*-forms result in positive density where *holo*-protein atoms occupy regions that are unoccupied in the *apo*-form, and negative density where *apo*-protein atoms occupy regions that are unoccupied in the *holo*-form. The standard interpretation of such a map is that negative density indicates positions of protein atoms before ligand binding, and positive density locates the same atoms after ligand binding. In regions where the two forms are identical, $\mathbf{F}_{holo} - \mathbf{F}_{apo}$ = 0, and the map is blank.

Structural Properties of Crystalline ALBP. A Ramachandran plot of the main chain dihedral angle Φ and Ψ is shown in Fig. 8.6.

In the refined model, 13 residues have positive Φ angles, 9 of which belong to glycine residues. There are 11 glycine residues in ALBP, all associated with good quality electron density.

Most of the residues having forbidden values of Φ and Ψ are glycines, represented by "+" in Fig. 8.6, while all other amino acids are represented

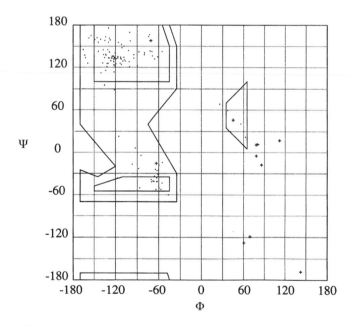

Figure 8.6 Ramachandran plot of the crystallographic model of ALBP. The main-chain torsional angle Φ (N–C$_\alpha$ bond) is plotted against Ψ (C$_\alpha$–C bond). The following symbols are used: (•) nonglycine residues; (+) glycine residues. The enclosed areas of the plot show sterically allowed angles.

by dots. Succeeding discussion reveals that these unusual conformations are also found in P2 and other members of this protein family, strengthening the argument that these conformations are not errors in the model, and suggesting that they might be important to structure and/or function in this family of proteins.

Plate 2 shows, at the end of refinement, the same section of map as in Plate 10. By comparing Plates 2 and 10, you can see that the map errors described above were eliminated, and that the map is a snug fit to a chemically, stereochemically, and conformationally realistic model.

IV. Summary

All crystallographic models are *not* equal. The noncrystallographer can assess model quality by carefully reading original publications of a macromolecular structure. The kind of reading and interpretation implied by my annotations in Section II are essential to wise use of models. Don't get me wrong; there is no attempt on the part of crystallographers to hide the limitations of models. On the contrary, refinement papers often represent almost heroic efforts to make plain what the final model says and leaves unsaid. These efforts are in vain if the reader does not understand them, or worse, never reads them. These efforts are often undercut by the simple power of the visual model. The brightly colored stereo views of a protein model, which are in fact more akin to cartoons than to molecules, endow the model with a concreteness that exceeds the intentions of the thoughtful crystallographer. It is impossible for the crystallographer, with vivid recall of the massive labor that produced the model, to forget its shortcomings. It is all too easy for users of the model to be unaware of them. It is also all too easy for the user to be unaware that, through temperature factors, occupancies, undetected parts of the protein, and unexplained density, crystallography reveals more than a single molecular model shows.

Even the highest-quality model does not explain itself. If I showed you a perfect model of a protein of unknown function, it is highly unlikely that you could tell me what it does, or even pinpoint the chemical groups critical to its action. Using a model to explain the properties and action of a protein means bringing the model to bear upon all the other available evidence. This involves gaining intimate knowledge of the model, a task roughly as complex as learning your way around a small city. In the next chapter, I will discuss the exploration of crystallographic models by computer graphics.

9 Tools for Studying Proteins

I. Introduction

There is an old line about a dog who is finally cured of chasing cars—when it catches one. What to do now? In this chapter, I will discuss what to do when you catch a protein. My main goal is to inform you about the tools available for studying protein models, and to suggest strategies for learning your way around the unfamiliar terrain of a new protein. I will begin with a very brief glimpse of the computations that underlie molecular graphics displays. Then I will take you on a tour of molecular modeling by detailing the features present on most modeling programs. Finally, I will briefly introduce other computational tools for studying and comparing proteins.

II. Computer models of molecules

A. *Two-dimensional images from coordinates*

Computer programs for molecular modeling provide an interactive, visual environment for displaying and exploring models. The fundamental operation of computer programs for studying molecules is producing realistic displays, convincing images of molecular models. Although the details of programming for graphics displays vary from one program (or programming language, or computer operating system) to another, they all produce an image according to the same geometric principles.

A display program uses a file of atomic coordinates to produce a drawing on the screen. Recall that a coordinate file contains a list of all atoms located by crystallographic analysis, with coordinates x, y, and z for each atom. When the model is first displayed, the coordinate system is shifted by the modeling program so that the origin is the center of the model. This origin is displayed at the center of the screen, becoming the origin of a new coordinate system, the screen coordinates, which I will designate x_s, y_s, and z_s. The x_s axis is displayed horizontally, y_s is vertical, and z_s is perpendicular to the computer screen. As the model is moved and rotated, the screen coordinates are continually updated.

The simplest molecular displays are stick models with lines connecting atoms, and atoms simply represented by vertices where lines meet. The stereo images in this book are of this type. It is easy to imagine a program that simply plots a point at each position (x_s, y_s, z_s) and connects the points with lines according to a set of instructions about connectivity of atoms in amino acids.

But the computer screen is two-dimensional. How does the computer plot in three dimensions? It doesn't; it plots a projection of the three-dimensional stick model. Mathematically, projecting the object into two dimensions involves some simple trigonometry, but graphically, projecting is even simpler. The program plots points on the screen at positions $(x_s, y_s, 0)$; in other words, the program does not employ the z_s coordinate in producing the display. This produces a projection of the molecule on the $x_s y_s$ plane of the screen coordinate system, which is the computer screen itself (see Fig. 9.1).

B. Into three dimensions: Basic modeling operations

The complexity of a protein model makes it essential to display it as a three-dimensional object and move it around (or move our viewpoint around within it). The first step in seeing the model in three dimensions is rotating it, which gives many three-dimensional cues and greatly improves our perception of it. Rotating the model to a new orientation entails calculating new coordinates for all the atoms and redisplaying by plotting on the screen according to the new (x_s, y_s) coordinates.

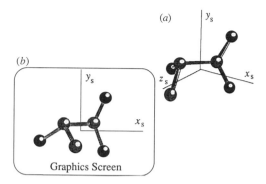

Figure 9.1 Geometry of projection. (*a*) Model viewed from off to the side of the screen coordinate system. Each atom is located by screen coordinates x_s, y_s, and z_s. (*b*) Model projected onto graphics screen. Each atom is displayed at position $(x_s, y_s, 0)$, producing a projection of the model onto the $x_s y_s$ plane, which is the plane of the graphics screen.

The arithmetic of rotation is fairly simple. Consider rotating the model by θ degrees around the x_s (horizontal) axis. It can be shown that this rotation transforms the coordinates of point $p[x_s(p), y_s(p), z_s(p)]$ to new coordinates $[x'_s(p), y'_s(p), z'_s(p)]$ according to these equations:

$$x'_s(p) = x_s(p)$$
$$y'_s(p) = y_s(p) \cdot \cos\theta - z_s(p) \cdot \sin\theta \qquad (9.1)$$
$$z'_s(p) = y_s(p) \cdot \sin\theta + z_s(p) \cdot \cos\theta$$

Notice that rotating the model about the x_s-axis does not alter the x_s coordinates, $x'_s(p) = x_s(p)$, but does change y_s and z_s. A similar set of equations provides for rotation about the y_s- or z_s-axis. If we instruct the computer to rotate the model around the x_s-axis by θ degrees, it responds by converting all coordinates (x_s, y_s, z_s) to new coordinates (x'_s, y'_s, z'_s) and plotting all points on the screen at the new positions $(x'_s, y'_s, 0)$. Most graphics programs allow so-called real-time rotation, in which the model appears to rotate continuously. This requires fast computation, because to produce what looks like smooth rotation, the computer must produce a new image about every 0.05 second. So in literally the blink of an eye, the computer increments θ by a small amount, recalculates coordinates of all displayed atoms, using

equations like (9.1), and redraws the screen image. Fast repetition of this
process gives the appearance of continuous rotation.

C. Three-dimensional display and perception

Complicated models become even more comprehensible if seen as three-dimensional (3D) objects even when not in motion. So graphics display programs provide some kind of full-time 3D display. This entails producing
two images like the stereo pairs used in this book, and presenting one image
to the left eye and the other to the right eye, which is the function of a stereo
viewer. The right-hand view is just like the left-hand view, except that it is
rotated about 5° about its y_s axis (clockwise, as viewed from above). Molecular modeling programs display the two images of a stereo pair side-by-
side on the screen, for viewing in the same manner as printed pairs.

   With proper hardware, modeling programs can produce full-screen 3D.
The technique entails flashing full-size left and right views alternately at
high speed. The viewer wears special glasses with liquid-crystal lenses that
alternate rapidly between opaque and transparent. When the left-eye view
is on the screen, the left lens is clear and the right lens is opaque. When the
screen switches to the right-eye view, the right lens becomes clear and the
left becomes opaque. In this manner, the left eye sees only the left view, and
the right eye sees only the right view. The alternation is fast, so switching
is undetectable, and a full-screen 3D model appears to hang suspended before the viewer.

   Both types of stereo presentation mimic the appearance of objects to our
two eyes, which produce images on the retina of objects seen from two
slightly different viewpoints. The two images are rotated about a vertical
axis located at the current focal point of the eyes. From the difference between the two images, called *binocular disparity*, we obtain information
about the relative depth of objects in our field of view. For viewers with
normal vision, two pictures with a 5° binocular disparity, each presented to
the proper eye, gives a convincing three-dimensional image.

   Unfortunately, a significant percentage of people cannot obtain depth information from binocular disparity alone. In any group of twenty students,
there is a good chance that one or more will not be able to see a 3D image
in printed stereo pairs. In the real world, we decode depth not only by binocular disparity but also by relative motion, size differences produced by
perspective, overlap of foreground objects over background objects, effects
of lighting, and other means. Printed stereo pairs provide only one type of

depth cue. Molecular modeling programs provide depth information to a wider range of viewers by providing depth cueing in the form of shading, perspective, and movement, as described below.

III. Touring a typical molecular modeling program

As you might infer from a brief description of the computing that underlies molecular graphics, the computer must be fast. Although programs abound for displaying simple molecules on common personal computers, they are used primarily to create still pictures for publication. Ordinary personal computers, and older mainframe computers as well, are too slow for exploring complex models. Most molecular modeling is done on graphics workstations: self-contained, high-speed computers designed especially for extremely fast computation and display. The fastest current models use so-called RISC (reduced-instruction-set computer) components. Of course, the evolution of computer technology is rapid, and personal computers employing this design are now appearing. Without doubt, your personal computer will soon give you useful molecular graphics, even for large molecules.

The basic operations of projecting and rotating a screen image of the molecular model underlie all molecular graphics programs. On these operations are built many tools for manipulating the display. These tools give viewers the feeling of actively exploring a concrete model. Now I will discuss tools commonly found in modeling programs. Although this is a general discussion of modeling tools, it is based primarily on my personal experience with the UCSD Molecular Modeling System (MMS), a sophisticated but inexpensive molecular graphics program developed by Steve Dempsey of the Department of Chemistry Computer Facility, University of California, San Diego. Illustrations cited in the following sections were made by photographing MMS displays on a Silicon Graphics Integrated Raster Image System (IRIS) workstation.

A. Importing and exporting coordinate files

As indicated above, image display and rotation requires rapid computing. Reading coordinates from a text file like a PDB (Protein Data Bank) file is slow, because every letter or number in the file must be translated into binary or machine language for the computer's internal processing. For this reason, graphics programs work with a machine-language version of the coordinate file, which can be read and recomputed faster than a text file. So the first step in exploring a model is usually converting the PDB file to binary form. Depending on the software, this may be done within the modeling program by specifying the text file for input and a name for the new binary file, or may be carried out by an auxiliary program that runs independently of the graphics program.

Often users produce revealing views of the model, and wish to use the coordinates in other programs, such as energy calculations or printing for publication. For this purpose, molecular modeling programs include routines for writing coordinate files in standard formats like PDB, using the current binary model coordinates as input.

B. Loading and saving models

Binary coordinate files used by graphics programs can be loaded and saved just like any other files. As the model is manipulated or altered, coordinates are updated. At any point, the user can save the current model in the binary form used by the program, or replace the current model with one saved earlier. To facilitate studying intermolecular interactions, modeling programs can handle several models at once, perhaps as many as 10 at one time. The models currently in memory can be viewed and manipulated individually or together. When they are saved, they retain their current relative orientations.

C. Viewing models

Standard viewing commands allow users to rotate the model around screen x_s-, y_s-, and z_s-axes, and move (translate) the model for centering on areas of interest. Viewers magnify the model by moving it along the z_s-axis toward the viewer, or by using a command that magnifies the image without changing coordinates by simply narrowing the viewing angle. "Clip" or "slab" commands simplify the display by eliminating foreground and background, producing a thin "slab" of displayed atoms.

Plate 11 shows two views of the small protein cytochrome $b5$,[1] an electron-transport protein containing an iron-heme prosthetic group (shown in

---

[1] F. S. Mathews, M. Levine, and P. Argos, *Nature* **233**, 15 (1971). Atomic coordinates (revised 1990) obtained from the Protein Data Bank (Chapter 7).

green with red iron at the center). In (a), you see the heme from behind its binding pocket, with much of the protein in the foreground. In (b), the foreground is clipped to give an unobstructed view of the heme and protein groups above and below it.

The program clips by displaying only those atoms whose current z_s-coordinates lie within a specified range, which is chosen visually by sliding front and rear clipping planes together until unwanted background and foreground atoms disappear. "Center" commands allow the user to select an atom to be made the center of the display. On selection by pointing to the atom and clicking a mouse, or by naming the atom, the program moves the model so as to center the desired atom on the screen and within the viewing slab.

The viewing commands also include selection of stereo or mono viewing, and offer various forms of depth cueing to improve depth perception, either by mimicking the effects of perspective (front of model larger than rear), shading (front of model brighter than rear), or rocking the model back and forth by a few degrees of rotation about the y_s-axis. In Plate 11, a menu on the left of the screen allows selection of many of the functions I have mentioned.

Finally, as an aid to adjusting the view, some viewing tools include a miniature image of the whole molecule as seen from above the computer, with lines marking the viewer's line of sight (screen z_s-axis), the clipping planes, and the viewing angle. This visual aid is also shown in Plate 11, on the right of the screen. The model is the multicolored cluster at the lower right. The long vertical green line represents the viewer's line of sight, the green "V" shows the viewing angle, and yellow horizontal lines show clipping planes. Only atoms between the clipping planes and within the viewing angle are included in the display. The near clipping plane is marked H ("hither") and the far plane Y ("yon"). In (a), the entire model lies between the clipping planes. In (b), the H plane is moved into the model to a position just in front of the heme binding pocket, eliminating foreground atoms from the display.

This image of "model space" provides clearer understanding of movements along the z_s-axis, and allows the viewer to find clipping planes, which are invisible in the display unless they are cutting the model. It is not unusual for novice viewers to accidentally move the model completely out of view and be unable to find it. Nothing is more disconcerting to the beginner than completely losing sight of the model. When the model disappears, it may be off to the side of the display, above or below the display, or still centered but outside the slab defined by clipping planes. In any event, the miniature image can often help viewers find the model and regain their bearings. As a last resort, there is usually a reset command, which brings model and clipping planes back to starting positions. The viewer pays a price for resetting, losing the sometimes considerable work of finding a particularly clear orientation for the model, centering on an area of interest, and clipping away obscuring parts of the structure.

D. Editing and labeling the display

A display of every atom in a protein is often forbidding and incomprehensible. Viewers are interested in some particular aspect of the structure, such as the active site or the path of the backbone chain, and may want to delete irrelevant parts of the model from the display. "Edit" commands allow viewers to turn atoms on and off. Atoms not on display continue to be affected by rotation and translation, so they are in their proper places when redisplayed. Viewers might eliminate specific atoms by pointing to them and clicking a mouse, or eliminate whole blocks of sequence by entering residue numbers. They may display only α-carbons to show the folding of the protein backbone (refer to Plate 4), or only the backbone and certain side chains to pinpoint specific types of interactions.

Editing requires knowledge of how atoms are named in the coordinate file, which is often, but not always, the same as PDB naming. (Refer to Chapter 7, Section VII.) Thus, viewers can produce an α-carbon-only model by limiting the display to atoms labeled CA. Each program has its own language for naming atoms, residues (by number or residue name), distinct chains in the model (like the α and β chains of hemoglobin), and distinct models. Viewers must master this language in order to edit displays efficiently. (See the next section, on coloring, for an example of how naming atoms can simplify complicated tasks.)

Even with an edited model, it is still easy for viewers to lose their bearings. "Label" commands attach labels to specified atoms, signifying element, residue number, or name. Labels float with the atom during subsequent viewing, making it easy to find landmarks in the model (see, for example, the labels of N- and C-termini in Plate 4).

E. Coloring

While you may think that color is merely an attractive luxury of more powerful computers, adding color to the display makes it dramatically more understandable. "Color" commands allow users to identify elements or specific residues, or to emphasize structural elements. Most programs allow atoms to be colored manually, by picking a color from a palette and then pointing to an atom and clicking the mouse. Alternatively, viewers color large selections of atoms by choosing a color and then naming atoms, residues, or sequence numbers. For example, a viewer can choose red, then specify atom name O to color all backbone carbonyl oxygens red; or enter all residue ranges listed as α-helical in the PDB file, and thus, in one step, color all α-helices red.

F. Measuring

Measurements are necessary in identifying interactions within and between molecules. In fact, noncovalent interactions like hydrogen bonds are defined by the presence of certain atoms at specified distances and bond angles from each other. In Plate 12, the yellow dotted line connects an oxygen atom (red) of a heme carboxyl group to the oxygen of a serine side chain. The distance between atoms, 2.61 Å, is displayed, and is approximately the distance expected for oxygen atoms involved in a O—H—O hydrogen bond. (Recall that crystallographic studies of proteins do not usually reveal hydrogen atoms.)

Modeling programs allow display of distances, bond angles, and dihedral angles between bonded and nonbonded atoms. These measurements float on dotted lines within the model (just like labels) and are active. they are continually updated as the model is changed, as described in the next section.

G. Exploring structural change

Modeling programs allow the viewer to explore the effects of various changes in the model, including conformational rotation, change in bond length or angle, and movement of fragments or separate chains. Used along with active measurements, these tools allow viewers to see whether side chains can move to new positions without colliding with other atoms, or to examine the range of possible movements of a side chain. In Plate 12, the side chain of the heme group is prepared for exploring conformational change. Bonds marked with curved arrows can be rotated by mouse movements, as indicated by a double-arrow mouse pointer at the lower right. The bond marked by the green curved arrow is currently active for rotation. As the user rotates atoms about this bond, the yellow dotted line will continue to follow the two oxygen atoms, and the measurement will be updated.

Like rotation and translation, changes of model conformation, bond angles, or bond lengths are reflected by changes in the coordinate file. The changes are tentative at first, while users explore various alterations of the model. After making changes, users have a choice of saving the changes, removing the changes, or resetting in order to explore again from the original starting point.

H. Exploring the molecular surface

Stick models of the type shown here are the simplest and fastest type of model to compute and display, because they represent the molecule with the smallest possible number of lines drawn on the screen. Stick models are relatively open, so the viewer can see through the outer regions of a complex molecule into the interior, or into the interface between models of interacting molecules. But when the viewer wants to explore atomic contacts, a model of the molecular surface is indispensable.

Published structure papers often contain impressive space-filling computer images of molecules, with simulated lighting and realistic shadows and reflections. These images require the computer to draw hundreds of thousands of multicolored lines, and so the computer cannot redraw such images fast enough for continuous movements. Some of these views require hours to draw just once. While such views show contacts between atomic surfaces, they are not practical for exploring the model interactively. They are used primarily as snapshots of particularly revealing views.

How then to study the surface interactively? The most common compromise is called a "dotted surface" (Plate 13), in which the program displays dots evenly spaced over the surface of the molecule. This image reveals the surface without obscuring the atoms within, and can be redrawn rapidly as the viewer manipulates the model. Several types of surfaces can be computed, each with its own potential uses. One type is the van der Waals surface, in which all dots lie at the van der Waals radius from the nearest atom, the same as the surface of space-filling models. This represents the surface of "contact" between nonbonded atoms. Any model manipulations in which van der Waals surfaces penetrate each other are sterically forbidden. Van der Waals surfaces show packing of structural elements with each other, but the display is complicated because all internal and external atomic surfaces are shown.

Another useful surface display is the *solvent-accessible surface*, which shows all parts of the molecule that can be reached by solvent (usually water) molecules. This display omits all internal atomic surfaces, including crevices that are open to the outside of the model, but too small for solvent to enter. Some modeling programs contain routines for calculating this surface, while others can take as input the results of surface calculations from widely available programs. Calculating the solvent-accessible surface entails simulating the movement of a sphere (called a *probe*) having the diameter of a solvent molecule over the entire model surface, and computing positions of evenly spaced dots wherever model and solvent come into contact.

Carrying out the same simulation that produces solvent-accessible surface displays, but locating the dots at the center of the probe molecule produces the *extended surface* of the model. This display is useful for studying intermolecular contacts. If the user brings two models together, one with extended surface displayed, the other as a simple stick model, the points of intermolecular contact are where the extended surface of one model touches the atom centers of the second model.

In Plate 13, the green dots represent the extended surface of the heme group in cytochrome $b5$. Because the dots are two atom diameters from heme atoms, they are at the same distance as the centers of protein atoms in contact with the heme. So contacts are easily detectable as vertices or tips of the protein model that just graze this extended surface display, such as two ring atoms of the phenylalanine side chain (red) at the top center of the heme surface. It is far easier to find specific interactions in this display than when both contacting surfaces are dotted.

The color of the displayed surface is usually the same as the color selected for the underlying atoms. In Plate 13, the surface of the heme carboxyl oxygen produces the red bulge at the back upper left of the heme. The red oxygen atom of serine touches this surface, displaying the same interaction shown in Plate 12. Alternatively, color can reflect surface charge (commonly, blue for positive, red for negative, with lighter colors for partial charges), or surface polarity (contrasting colors for hydrophobic and hydrophilic regions). These displays facilitate finding regions of the model to which ligands of specified chemical properties are likely to bind.

I. Exploring intermolecular interactions: Multiple models

Formulating proposed mechanisms of protein action requires investigating how proteins interact with ligands of all kinds, including other proteins. Molecular modeling programs allow the user to display and manipulate several models, either individually or together. Tools for this purpose usually allow all of the same operations as the "viewing" tool, but permits selection of models affected by the operations. In "docking" experiments (a term taken from satellite docking in the space program), one model can be held still while another is moved into possible positions for intermolecular interaction. Labeling, measurement, and surface tools are used simultaneously during docking to assure that the proposed interactions are chemically realistic. Some programs include computational docking, in which the computer searches for optimal interaction, usually from a user-specified starting point.

J. Displaying crystal packing

Many molecular modeling programs include the capacity to display models of the entire unit cell. All the program needs as input is a set of coordinates for one molecule, the unit cell dimensions, and a list of equivalent positions for the crystal space group. The user can display one cell or clusters (say, $2 \times 2 \times 2$) of cells. The resulting images, particularly when teamed with surface displays, reveal crystal packing interactions, allowing the user to see which parts of the crystallographic model might be altered by packing, and might thus be different from the solution structure. For examples of crystal packing displays, see Plates 5 and 6. Unit-cell tools usually allow the user to turn equivalent positions on and off individually, making them useful for teaching the topics of equivalent positions and symmetry.

K. Building models from scratch

As well as taking coordinate files as inputs, modeling programs allow the user to build peptides to specification, and to change amino-acid residues within a model. To build new models, users select amino acids from a palette or list, and direct the program to link the residues into chains. Users specify conformation for the backbone by entering backbone angles Φ and Ψ, by selecting a common secondary structure, or by using the tools described above for exploring structural change. Model-builder tools are excellent for making illustrations of common structural elements like helices, sheets, and turns. The same tools are used to replace one or more side chains in a model with side chains of different amino acids, and thus explore the local structural effects of mutation.

IV. Other tools for studying structure

It is beyond the scope of this little book to cover all the tools available for studying protein structure. I will conclude by listing and briefly describing additional tools, especially ones used in conjunction with modeling on graphics systems.

A. Tools for structure analysis

In addition to molecular graphics, a complete package of tools for studying protein structure includes many accessory programs for routine structure analysis. The chores executed by such programs include the following:

- Calculating Φ and Ψ angles and using the results to identify elements of secondary structure.
- Using distance and angle criteria to search for hydrogen bonds and salt links, and producing a list of such interactions.
- Comparing homologous structures by least-squares superposition of one protein backbone on another. The result is a new coordinate set for one model that best superimposes it on the other model. The superimposed models can be viewed by molecular graphics. (I used such a tool to compare the x-ray and NMR structures of thioredoxin in Plate 4.)
- Calculating surface electric fields, which can be displayed in graphics programs to reveal regions that would attract molecules of opposite charge, or to show expected direction of movement of charged ligands.

B. Tools for modeling protein action

The crystallographic model is used as a starting point for further improvement of the model by energy minimization and for simulations of molecular motion. Additional insight into molecular function can be obtained by calculating charge densities and bond properties by molecular orbital theory. For small molecules, some of these calculations can be done "on the fly" as part of modeling. For the more complex computations, and for larger molecules, such calculations are done outside the graphics program, often as separate tasks on computers whose forte is "number crunching" rather than graphics.

V. A final note

Molecular modeling, just like printed pictures of protein models, endows models with the concreteness of everyday objects. While exploring models, viewers can easily forget the difficult and indirect manner by which molecular images are obtained. I wrote this book in hopes of providing an intellectually satisfying understanding of the origin of crystallographic models. I also hope to encourage readers to explore the many models now available, but to approach them with full awareness of what is known and what is unknown about the molecules under study. Just as good literature depicts characters and situations in a manner that is "true to life," a sound crystallographic model depicts a molecule in a manner that is true to the crystallographic data. But just as real life is more multifarious than the events, settings, and characters of literature, not all aspects of molecular truth (or even crystallographic truth) are reflected in the colorful model floating before us on the computer screen. The user must probe deeper, into the crystallographer's esoterica, to know just where the graphics depiction is not faithful to the crystallographic data. The user must probe further still, into the wider literature on the molecule, to know whether the crystallographic model is faithful to other evidence about structure and action. The conversation between structural models and evidence on all sides continually improves models as depictions of molecules.

Another possible effect of this book may be to stimulate your interest in crystallography itself. You may be wondering where to go from here. As your next step toward a truly rigorous understanding of crystallography, I suggest *X-ray Structure Determination: A Practical Guide*, 2nd edition, by George H. Stout and Lyle H. Jensen (John Wiley and Sons, Inc., 1989), or *Crystal Structure Analysis: A Primer*, 2nd edition, by Jenny Pickworth Glusker and Kenneth N. Trueblood (Oxford University Press, 1985).

Index

Absences, systematic, 61, 80, 97–98, 170
Absolute configuration, *see* Hand problem
Absorption edge, 63, 118
Adipocyte lipid-binding protein (ALBP), 8,
    32, 88, 142, 150, 162, 163, 166, 183
Alignment
    camera, 67, 73, 75
    protein sequence with electron-density map,
        143
Amplitude
    structure factor, 27, 94
    wave, 20, 84
Annular screen, 75
Anomalous dispersion, *see* Anomalous
        scattering
Anomalous scattering, 118–124
    determining hand, 123
    phasing, 120
Area detector, 68, 72, 77, 168, 169
Asymmetric unit, 40, 41, 59, 61, 170
Atomic coordinates, *see* Coordinates, atomic
Axes
    coordinate, 44, 51
    symmetry, 60
        rotation, 59
        screw, 60, 61, 97, 117, 170
    unit cell, 44, 78

B, *see* Temperature factor
Beam stop, 67, 72
Bias, model, 138, 140
Bond angle, 27, 148, 152, 158, 172, 180, 193
Bond length, 27, 148, 152, 158, 172, 180

Bragg's law, 43, 48–50, 92
    real space, 48–50
    reciprocal space, 52–55

Camera, 63, 67, 69
    alignment, 67, 75
    oscillation, 72, 75
    precession, 72, 73
    rotation, 72, 75
Cell constants, *see* Cell parameters
Cell parameters, 14, 44, 57, 78–80, 169, 170,
        171
Clipping, 191
Collimator, 67, 72, 168
Complex number, 85, 102
Complex vector, 102, 103–105
Convergence
    criteria, 151–152, 158, 162
    to final model, 137, 149
    radius, 147
Coordinates
    atomic, 9, 91, 142
        in modeling, 185
    Patterson space, 114, 116
    real space, 18
    reciprocal space, 18
Coordinate systems, 18
Crystal, 8
    density, 40, 137
    growing, 9, 35–40
    orienting, 63, 69, 75
    properties, 29–33
    seed, 37, 168

Crystal (*continued*)
    selecting, 32, 40
    size, 29, 31
    twinned, 29, 40
    water content, 31, 32–33, 35, 40, 137, 170
Crystallization, 9, *see also* Crystal, growing
Cytochrome *b*5, 190

Data
    collecting, 10–11, 17, 61–81, 108, 168,
        172
        symmetry and strategy, 80
    postrefining, 77
    scaling, 77, 169
Density, crystal, 40, 137
Depth cueing, 189
Derivative, heavy-atom, 35, 38, 107–109,
    135
    phases from, 109–113
Difference Fourier, 124, 140, 180
Difference map, 181
Difference Patterson, 115
Diffraction, 6, 7, 10, 12–17, 24
    geometric principles, 43–61
Diffraction pattern, 11, 13–17, 18, 95
Diffractometer, 69, 71–72, 78, 123, 168, 173
Direct lattice, *see* Lattice, real
Direct methods, 124
Disorder, 162, 177
Dispersion, anomalous, *see* Anomalous
    scattering
DNA, 1
Docking experiments, 196

Electron density, 24, 83, 88
    from measured reflections, 27, 94, 107
    from structure factors, 25–27, 93
    unexplained, 165, 166
Electron-density map, 24–25, 27, 135, 176
    interpreting, *see* Map fitting
Energy minimization, 148, 197, *see also*
        Energy refinement
Energy refinement, 148, 174, 175
Equivalent positions, 60, 97, 117, 171, 196
Extinctions, *see* Absences, systematic

F_{ooo}, 135
Film, 11, 64, 67, 73, 77

Fourier analysis, 89, 90
Fourier series, 22, 84–88, 90, 95
    exponential form, 86
    trigonometric form, 85
Fourier synthesis, 22, 90
Fourier transform, 26, 83, 88–90, 91, 93, 133
Friedel pairs, 104, 118–123
Friedel's law, 80, 104, 118

Goniometer head, 69, 71, 73
Goniostat, 71, 72
Graphics workstation, 189

Hand problem, 118, 123–124
Hanging-drop method, 36, 168
Harker section, 117
Heavy-atom derivative, *see* Derivative,
    heavy-atom
Heavy-atom method, 107–118
Hydrogen bond, 1, 31, 149, 177, 193, 197

Indices, 18, 45–47, 50, 75, 91, 94, 95
Intensity, reflection, 11, 14, 16, 50, 77, 81,
    99
Isomorphous replacement, 107, 108, 113, 124
Iteration, 96, 129, 131, 133, 151, 176

K_α, 63–64, 123
K_β, 63–64, 118
    filtering, 63

Lattice, 9, 13
    real, 14
    reciprocal, 14, 50–52, 73
    types, 58
Laue group, 80
Least squares, 142, 144–145, 168, 169, 175,
    176, 197
    in structure refinement, 132, 145–149, 163
Limiting sphere, 56
Local minima, 147, 149
Luzzati plot, 161, 180

Map fitting, 27, 139, 142–143, 149
Microscope, 6

Model
  limitations, 3, 158–166, 183
  sharing, 142, 153–155
Molecular dynamics, 149, 164
Molecular modeling, 185–197
Molecular replacement, 125–129, 172, 173,
    177
  orientation search, 127, 128, 174
  Patterson correlation refinement, 175
  translation search, 127, 128, 175
Molecular weight, 40, 41
Mosaic spread of reflections, 31, 77
Mosaicity, 31, 77

Optical diffraction patterns, 13
Oscillation photograph, 75, 78

Parameters, unit cell, 14, 44, 57, 78–80, 169,
    170, 171
Patterson
  atom, 116
  correlation refinement, 175, 178, 179
  function, 114, 118, 127, 128
  map, 115, 118, 127, 128, 174, 175
Periodic function, 20–25, 84, 86, 90, 95
Phase angle, 104, 106, 107, 111
Phase probability, 113, 114, 123, 151
Phase problem, 27, 35, 94, 107, 124
Phases
  anomalous scattering, 118–124
  extending, 138
  heavy atom, 107–118
  improving, 96, 129, 132, 136
  molecular replacement, 125, 129, 172
Phasing model, 125, 132, 172
  isomorphous, 125
  nonisomorphous, 126–129
Postrefinement of data, 77
Precession camera, 72, 73
Precession photograph, 11, 73, 75, 78, 81, 108
Protein Data Bank, 142, 150, 153–155, 164,
    173
  file contents, 154

R-factor, 128, 152, 158, 160, 161, 169, 172,
    175
Radius of convergence, 147
Ramachandran angles, 159

Ramachandran diagram, 159, 176, 182
Real lattice, 14
Real space, 18, 90, 151
Reciprocal lattice, 14, 50–52, 73
Reciprocal space, 18, 50, 52, 144, 151
Refinement, 149, 151, 175
  energy, 174, 175
  least squares, 132, 144, 145, 149
Reflections
  number of, 56–57, 173
  unique, 80–81, 173
Regularization, 142, 151
Resolution, 18, 57, 64, 95, 139, 150, 164,
    169, 170, 173
  and precision of atomic positions, 160–162
RISC computers, 189
Rotation axis, 59
Rotation function, 174, 175, 179

Scaling, 77, 169
Scattering factors, atomic, 91, 147
Scintillation counter, 68, 71
Screw axis, 60, 61, 97, 117, 170
Space group, 58–61, 173, 196
  determining, 80, 97, 170
Sphere of reflection, 55–57, 63, 64, 74, 75,
    78, 169
Standard deviation, 144, 145
Stereo diagrams, 188
Stereo viewing, 1
Structure factor, 23–24, 25–27, 91–93
  amplitude, 27, 94
  atomic, 91
  as complex vector, 102–105
  computing from model, 96, 132, 139, 174
Surface, molecular, 194–196
  extended, 195
  solvent-accessible, 194
  van der Waals, 24, 194
Symmetry
  determining, 58, 61, 97–98, 170
  noncrystallographic, 138
Synchrotron radiation, 64, 65, 118
Systematic absences, see Absences, systematic

Temperature factor, 146, 147, 162–163, 165,
    176, 177, 180
  anisotropic, 163
  isotropic, 163

Thermal motion, 162, 177
Three-dimensional display, 185–189
Twinned crystal, 29, 40

Unit cell, 9, 18, 44, 49
   graphics display, 196
   parameters, 14, 57, 78, 169, 170, 171
   reciprocal, 51
   symmetry, 58–61

Vapor diffusion, 36
Vector
   complex, 102, 103, 105
   Patterson, 115
Vibration, atomic, 162–164

Water in crystals, 31, 32–33, 35, 40, 137, 170
Wave amplitude, 20, 84
Wave equation, 20–25, 84
Wavelength, 7, 20
   x-ray, 63–64

X-ray, 7, 63–67
   absorption edge, 63, 118
   camera
      oscillation, 75
      precession, 73
      rotation, 75
   damage to crystal, 35
   detectors, 67–69
   filter, 63
   health hazards, 67
   safety precautions, 67
   sources, 63–67
   synchrotron, 65
   tube
      cathode, 64
      rotating anode, 64
   wavelength, 7, 63–64
XPLOR, 173, 174

Zif268, 1
Zinc-finger protein, 1